T0351939

Prepare
and
Protect
Safer Behaviors in Laboratories
and Clinical Containment Settings

Prepare
and
Protect

Safer Behaviors in Laboratories and Clinical Containment Settings

Sean G. Kaufman MPH, CHES, CPH, CIC

Certified MBTI Personality Assessment Provider

International Federation of Biosafety Associations (IFBA)
  Certified Professional

Safer Behaviors, Woodstock, Georgia

ASM
PRESS

ASM Press
Washington, DC

WILEY

Editorial Correspondence: ASM Press, 1752 N Street, NW, Washington, DC 20036-2904, USA
Registered Offices: John Wiley & Sons, Inc., 111 River Street, Hoboken, NJ 07030, USA

For details of our global editorial offices, customer services, and more information about Wiley products, visit us at www.wiley.com.

Wiley also publishes its books in a variety of electronic formats and by print-on-demand. Some content that appears in standard print versions of this book may not be available in other formats.

Library of Congress Cataloging-in-Publication Data

Names: Kaufman, Sean G., author.
Title: Prepare and protect : safer behaviors in laboratories and clinical containment
    settings / Sean G. Kaufman, Safer Behaviors, Woodstock, Georgia.
Description: Washington, DC : American Society for Microbiology ; Hoboken, NJ :
    Wiley, [2020] | Series: ASM books | Includes index.
Identifiers: LCCN 2019051486 (print) | LCCN 2019051487 (ebook) |
    ISBN 9781683670148 (paperback) | ISBN 9781683670179 (adobe pdf)
Subjects: LCSH: Biological laboratories–United States–Safety measures. |
    Medical laboratories–United States–Safety measures. | Laboratory infections–United
    States–Prevention.
Classification: LCC QH323.2.K38 2020 (print) | LCC QH323.2 (ebook) |
    DDC 570.72–dc23
LC record available at https://lccn.loc.gov/2019051486
LC ebook record available at https://lccn.loc.gov/2019051487

Cover images: (front) Colorized scanning electron micrograph of filamentous Ebola virus particles (blue) budding from a chronically infected VERO E6 cell (yelllow-green). Image captured and color-enhanced at the NIAID Integrated Research Facility in Ft. Detrick, Maryland. Credit: NIAID. (back) Courtesy Bethany Bandera https://www.bethanybandera.com/ Cover design by Susan Brown Schmidler

Printed in the United States of America

V10017179_012820

*This book is dedicated to those serving on the front lines
of infectious disease—the infectious disease pioneers. May you remain
safe as you serve others with your courage and determination in
fighting against infectious diseases.*

Contents

Foreword ix
Preface xi
About the Author xv

Chapter 1
Infectious Disease Pioneers 2
Biosafety in the First Person: *The Loss of Beth Griffin* by Caryl Griffin

Chapter 2
Understanding Containment 8
Biosafety in the First Person: *My First Patient* by Dottie Cappola-Vojak

Chapter 3
Biological Risk Mitigation 16
Biosafety in the First Person: *The Lab Mom* by Carrie Anglewicz

Chapter 4
Human Risk Factors 26
Biosafety in the First Person: *Malcolm Was My Friend*
by Joseph Kanabrocki

Chapter 5
Four Primary Controls of Safety 40
Biosafety in the First Person: *The Dancer of Biosafety* by Jim Welch

Chapter 6
Understanding Human Behavior 64
Biosafety in the First Person: *It Could Have Been HIV* by Henry Mathews

Chapter 7
The Containment Philosophy 76
Biosafety in the First Person: *I'd Do It All Over Again* by Lee Alderman

Chapter 8
Plans + Behaviors = Outcomes   90
Biosafety in the First Person: *Biosafety Found Me* by Anthony (AJ) Troiano

Chapter 9
Intrinsic Safety   102
Biosafety in the First Person: *The Bruises of Biosafety* by Sarah Ziegler

Chapter 10
Building a One-Safe Culture   122
Biosafety in the First Person: *An Incredible Journey* by Robert Hawley

Chapter 11
Emergency Preparedness and Response to Biological Risks   142
Biosafety in the First Person: *We All Have a Bucket* by Tim Trevan

Chapter 12
Standard Operating Behavior   164
Biosafety in the First Person: *The Biosafety Profession: An Unexpected Journey* by Joe Kozlovac

Chapter 13
Effective Training Strategies   180
Biosafety in the First Person: *She Was One of Us!* by Karen Byers

Chapter 14
The Beaking Method   192
Biosafety in the First Person: *Representing the Profession of Biosafety* by Ed Stygar

Chapter 15
Safety Surveillance Programs   204
Biosafety in the First Person: *Experience in Years* by Mike Pentella

Chapter 16
Responsible Leadership   220
Biosafety in the First Person: *The Road Less Traveled into Biosafety* by Bob Ellis

Image Credits   237
Index   239

Foreword

Sometimes you meet the right person at exactly the right time. Maybe you didn't even know in that moment how significant their impact might be; maybe you didn't realize how much you needed what they had to teach you. In late July of 2014, I met Sean Kaufman.

Since 2002, our facility has had a team of nurses, laboratorians, and physicians who are available to staff an inpatient hospital room in case a CDC worker were to be inadvertently exposed, either in the lab or in the field, to an infectious agent. We anticipated caring for, if necessary, an essentially well person while we observed them for the signs and symptoms of whatever contagious thing they might have been exposed to. We followed globally when there was an outbreak that would make the news: Marburg virus or dengue fever or Crimean-Congo hemorrhagic fever, or any other exotic-sounding, scary disease, whenever there was a possibility that a CDC worker might get a needlestick or be scratched by a bat or monkey or break a vial of blood. Our team waited, quietly and patiently, for these occasions that did not come. The CDC has an amazing safety record! We collected our on-call pay ($3.50/hour), for weeks and months and years on end, with barely a whiff of trouble.

And then, in 2014, we got the call. We knew an Ebola outbreak was raging along the western edge of Africa. We had read about the World Health Organization (WHO), Doctors Without Borders (MSF), and other relief organizations struggling to contain the outbreak and care for patients in makeshift hospitals across three countries. But our call was not from the CDC, it was from the State Department. Would we be willing to care for an American doctor from Liberia ill with Ebola virus disease? Were we ready? Could we be ready in a few short days?

And that's when I met Sean Kaufman. He jumped into a unit he did not know, into a team of people from across our health care system whom he did not know, to help review a strategy for safety to care for one of the world's deadliest diseases. And like a firefighter who enters a smoke-filled room to lead you to safety with a flashlight and a steady, guiding hand,

Sean came in and helped us make sure that we could be safe doing a job we had been trained for, but had never had to put into practice.

Because it is not practice alone that makes perfect. It is a watchful eye, it is a demonstration of competency, it's believing in each other as teammates, as family, and in a commitment to the safety of each person on that team, in that family. From Sean we were reminded of the importance of donning and doffing our personal protective equipment in a precise manner. He helped reinforce that our own behavior in that PPE, our attention to detail, our willingness to be accountable and hold others accountable, was what could keep us safe. And only by being safe ourselves could we keep our facility and our community, our friends and family, safe as well.

If you ever need to be led to safety, I hope you meet Sean Kaufman. I am grateful I did. I am grateful for the lessons that still resonate in my approach to patient care. Sean helped to instill in our team both the confidence to care for these precious patients and the weighty responsibility that comes with being on a team where there can be no mistakes, no weak links in the chain. He instilled in me a deeply and still held conviction that I cannot do this alone. That my safety is in the hands of my coworkers, and theirs in mine. To trust them, to believe in the processes and protocols, the practice and the perseverance of each member of this family.

Sometimes, you just meet the right person at the right time. Thank you, my friend.

Jill Morgan, RN
Emory Hospital, Atlanta, Georgia

Preface

Florence Nightingale said, "The first requirement of a hospital is that it should do the sick no harm." This statement served as the backbone of infection control for health care settings. I don't believe any scientist (of the right mind) has ever come to work with the goal of killing themselves or those with whom they work. Historically, there are several examples of scientists doing things to protect themselves when working in biological laboratories. However, it is my opinion the profession of biosafety formally started within the United States in 1941 when Secretary of War Henry Stimson stated, "Because of the dangers that might confront this country from potential enemies employing what may be broadly described as biological warfare, it seems advisable that investigations be initiated to survey the present situation and the future possibilities." In 1942, George Merck established the War Reserve Services under the stated premise, "There is but one logical course to pursue, namely, to study the possibilities of such warfare from every angle, make every preparation for reducing its effectiveness, and thereby reduce the likelihood of its use."

As I have traveled the world, I have witnessed vast differences between biosafety levels from country to country. Those countries that at one time or another had offensive and defensive biological weapons research programs are much further along in biosafety than those that did not have such programs. Principles, practices, and strategies aimed at keeping scientists alive while developing biological weapons were developed, to protect the scientists and proprietary information gained during these scientific experiments.

Like many others, I stand on the shoulders of giants. This is because many of my mentors were ones who took lessons learned in the age of biological weapons and generalized them to the world. The publication of the first edition of the *Biosafety in Microbiological and Biomedical Laboratories* (BMBL) in 1984 was foundational for formalizing the practice of biosafety and generalizing it to the outside world. The 6th edition of the BMBL will be published soon.

As a behaviorist, I often run into something called the Semmelweis Reflex, a metaphor for the reflex-like tendency to reject new evidence or new knowledge because it contradicts established norms, beliefs, or paradigms. Dr. Ignaz Semmelweis made a remarkable observation about the impact of a doctor's contaminated hands infecting women during childbirth. His hand-washing proposal was contrary to the beliefs and practices at the time. Dr. Semmelweis's ideas were rejected and ridiculed. He died alone in a mental institution, only to be acknowledged many years later for a discovery we still teach today—wash your hands with soap and water! Please consider reading this book with an open mind, as much of it may be contrary to what exists today.

Working with infectious diseases in clinical and laboratory environments has inherent risks. The World Health Organization defines inherent risk as risk associated with laboratory activities or procedures that are conducted in the absence of mitigation measures or controls. I believe both the professions of biosafety and infection control have done an outstanding job controlling inherent risks with engineering, personal protective equipment, and standard operating procedures. However, there is a risk that remains after these carefully measured controls are put into place—defined as residual risk.

I believe it is within residual risk that human risk factors exist. These inter- and intrapersonal skills—how a person interacts with the infectious disease, others they work with, and themselves—pose a risk which is not traditionally covered in health care or laboratory settings today. Further, there is the concept of "culture," which is the blend of the people, the agent, and the environment. We are just beginning to understand the challenges in developing and sustaining a culture of safety within an organization.

Although it is true that I have never officially served as a biosafety officer, I offer this book with the hope of promoting biosafety and addressing serious challenges. Today, biosafety professionals serve organizations, minimizing risks associated with handling biological agents in laboratory settings. I believe my role in biosafety is a unique one and very different from the activities of a biosafety professional.

While biosafety professionals work in the trenches (and trust me—this is very important work), I am using my gifts to inject excitement into safety, primarily around infectious diseases. I teach safer behaviors, and though I may not be able to provide a comprehensive essay on microbiology, I can discuss safer behavioral practices and explain human risk factors that are absolutely critical in the application of biosafety.

I have spent the last 15 years of my life in biosafety: motivating, inspiring, changing, and increasing awareness in biosafety. I am a promoter of biosafety, and since the 2014 Ebola outbreak, I am also a promoter of a term I coined, *clinical containment*. This book can be used to teach components of biosafety as well as to introduce the concept of biosafety to the health care industry. Health care staff must learn new ways of practicing infection control because what they are doing still has weaknesses. More illnesses can be prevented by combining the professions of biosafety and infection control with simple containment strategies.

Safety is not boring, because it centers on human behavior. In terms of safety, understanding human behavior is as important as understanding microbiology. It is my hope that this book will find its way into the hands of scientists, nurses, doctors, and health care and biosafety professionals. I hope this book increases your awareness and understanding of how behavior serves as the bridge between safety plans and safer outcomes.

I must acknowledge how thankful I am for the blessings I have received in this life. It started with being born to a mother and father who sacrificed a great deal to provide a life with so many memories. A brother who has always been there. Three beautiful children, and my wife Jacqueline, who has been my rock. I give all credit and praise to God—always have and always will.

Sean G. Kaufman

About the Author

I was born in San Diego, California. My father, a United States Marine Corps Officer (retired), is nothing more than a big teddy bear. My mother, short in stature, embodies a pit bull-like personality. She is extremely loving and capable of doing anything; as a family we learned to never get in her way. I also have a younger brother who was my best friend growing up. My father once said to me, "Son, for every action there is a reaction. If you don't like the reaction, change your action until you get the reaction you want. Then, go forward, as you have life figured out."

I was never the best student. Focused on my strength in sports, I was able to get into a school both my grandmother and father had attended—San Diego State University. When I first entered the University, I believed I wanted to be an elementary school teacher. I coasted through my first semester and then joined a fraternity. From that point on, school was never about academia, it was about where the next party was.

I will never forget the day I opened the mailbox and received the letter letting me know that I had been academically disqualified as an incoming junior. I truly did not know what to do. My parents were going to kill me. So, dressed in an old suit jacket, I drove myself across campus to the office of the director responsible for dismissing me from school. Once there, I ran right past the secretary and directly to Dr. Cathie Atkins, whom I refer to as one of the first angels of my life. I said to Dr. Atkins, "I just got this letter, and I need a second chance. Please!" After waiving her secretary Gloria off, Dr. Atkins looked at me and said, "What makes you think you deserve a second chance?" I looked at her and said, "I messed up. Lost my way. Please, I need your help."

She requested that I work with her all summer and told me that my second chance would be considered as I did so. I said that I had a job and other obligations. She demanded I resign from the job and the fraternity, which I did immediately. By the end of the summer, she called me into her office and changed my career path forever. "Sean, I notice you are not doing well in most of your classes, but you have straight As in all of your public health courses. I have enrolled you in a semester of public health courses, and we will see what you make of this second chance."

I never looked back, getting straight As and an occasional B for the rest of my undergraduate career. I brought Dr. Atkins flowers at the end of every subsequent semester, along with my report card, to say thank you for the second chance.

The last summer of my undergraduate career, I was summoned to her office. On hearing that I didn't know what my plans were after graduation, she asked if I had considered going to graduate school for public health. I responded with hesitation because of my past academic performance. She smiled and told me that the last 60 credits are the only ones considered. She suggested I take the GRE and work for Dr. Mel Hovell, who would eventually become the chair of my thesis committee.

I could go on about everything Dr. Atkins did for me but I think the greatest gift was one she gave to my father. She had nominated me for the growth and achievement award, which meant that I got to sit on stage with her during the graduation ceremony. My family was in attendance and began looking around for me among the throngs of students in the graduate seating area, but they could not find me. Dr. Atkins then stepped up to the microphone and said, "I remember the day I first met Sean Kaufman. He rushed into my office, past my secretary, and said he needed a second chance. I asked him why I should give him one and was surprised by his answer. Most of the time students blame teachers or something else but Sean didn't. He blamed himself; at that very moment I knew he deserved a second chance." From my seat on the stage, I saw my father cry, which in turn made me cry. I will never forget that day and will never be able to thank Dr. Atkins enough for everything she did. She is the reason I entered the School of Public Health and earned my MPH, and is part of why I am where I am today and writing this book.

There are many more qualified individuals than myself who are hard at work, right on the frontlines of this topic, and yet here I am sitting behind the computer with an opportunity to share the knowledge I have gained during my years working in this field.

I first fell in love with the field of infectious diseases through HIV. It is hard to even describe the stigma surrounding HIV, and very few people could understand my fascination with this virus; my parents thought I was either gay or sick with the disease. The fact that a virus could enter your body, recreate your RNA to include itself in your DNA, reproduce without your body knowing, and destroy your immune system blew my mind. What magnified my fascination was the hysteria surrounding HIV at the time; the virus was not yet well understood and, having no cure, was considered a death sentence.

My professional public health career started with HIV testing. I facilitated pretest sessions with patients, explaining what the test was and what the results meant and answering any questions. I would order the test, and the patient would make an appointment with me to get their results. It was a privilege to work with these patients, and also a great responsibility because when a result came back positive, the tone with which I delivered the news could have a powerful effect on the patient.

During this period of my career, I spent time with those who had lost loved ones, were losing loved ones, and would one day lose themselves to this virus. Even now, the memory of all those who gave their time and used their lives to raise awareness and change behaviors, with the hopes of preventing future HIV-related deaths, fills me with emotion.

The only reason I can think of that someone working in public health would ever move from San Diego, California, to Atlanta, Georgia, is to work at the Centers for Disease Control and Prevention (CDC). Following my move, I remember being in awe; some nights I could not even sleep because of how excited I was to be working at this pinnacle of public health. Things started slow there. The first protease inhibitors (medication) were approved in 1995, making HIV in the United States more of a chronic condition rather than an acute one. Furthermore, with a drug that controlled viral loads of HIV, the need for behaviorists decreased. I quickly transitioned to waterborne disease, where I worked with Dr. Michael Beach.

Dr. Beach is one of the most enthusiastic men I know. He loves parasites! During my stint in the Division of Parasitic Diseases, we developed the Healthy Swimming project. What fascinated me most during my time in this position was the behaviors I witnessed. If I were to have you and 20 people bathe together in a tub and then ask you to take the water that you bathed in and swish it around in your mouth, would you do it? Yet millions of people will jump into a swimming pool or lake and do exactly that! They swish with water that has touched perianal surfaces, that contains various bodily fluids and excretions, among numerous other things that could cause very serious illnesses.

It was a fun job, but things changed significantly following September 11, 2001. When the September 11th terrorist attacks occurred, I requested and was granted an opportunity to work within the emergency communications center. It was not long before I was deployed to Trenton, New Jersey, to work with postal employees who were exposed to *Bacillus anthracis* spores, the causative agent of anthrax, during what is now referred to as the 2001 anthrax attacks (or Amerithrax). I saw firsthand with both HIV and anthrax that behaviors around infectious disease outbreaks tended to be very similar, regardless of differences in the agents themselves. I met Suzanne Miro and worked with Jonathan King (who was a fellow at that time and is now a PhD working at the World Health Organization) and Paul Abamonte, and together we had a chance to meet Norma Wallace and her son, a survivor of inhalational anthrax who was featured in *People* magazine. When I returned to CDC, my career had changed.

Subsequently, I responded to outbreaks of West Nile Virus in Louisiana and severe acute respiratory syndrome (SARS) within the quarantine station at Los Angeles International Airport. I also trained with the Federal Emergency Management Agency at Mount Weather and in Anniston, Alabama. It was there that I met David Miller and Harvey Holmes, who would introduce me to David Bressler and Besty Weirich. Bressler and Weirich gave me my first introduction to biological laboratories and an understanding of what role these play during a bioterrorism event.

The next major influence on my life was a meeting that my colleague and friend Ellen Whitney set up with Dr. Ruth Berkelman. Ruth is an amazing woman; without her support, I would not be where I am today. After talking with Ruth for only a short time she said, "It is not a matter of if, it is a matter of when you will come work with me." I turned in my resignation notice at CDC and started working at Emory University in January of 2004.

Very soon after I started at Emory, Ruth entered my office and told me they had just received money to build a mock biosafety level 4 laboratory (BSL4). I thought it was great they had received their funding, but I wondered why they were building a mock lab and, more so, why she was talking to me about it. She kindly reminded me that I had a background in behavior and infectious diseases. She explained that the need for high-containment laboratory biodefense research was on the rise, but there was no capable workforce in place. We could use this mock lab to prepare and train exactly such a workforce.

During this project I met and worked with Lee Alderman, Henry Mathews, Rich Henkel, David Bressler, Betsy Weirich, and Peter Jahrling, all of whom had profound impacts on my biosafety career development. I jumped right in and started training groups in this mock BSL4. As we did more and more trainings, the need for and reputation of our program grew. This was also in no small part due to Ruth Berkelman's support. She never wavered in her commitment to the project!

The Emory University Science and Safety Training Program, as it would come to be called, took me around the world, delivering biosafety training programs on almost every continent and working with thousands of individuals. After 10 years, the training center grant had run out and the building that housed our laboratory was supposedly to be demolished (though it still stands today). At the time, Patty Olinger was the Director of Environmental Health Services at Emory. Patty offered me a part-time job working at the university, allowing me to retain my health insurance and continue serving as the CEO and Founding Partner of Safer Behaviors.

On July 31, 2014, I visited the Emory University isolation unit to determine whether they were ready to treat an Ebola-infected patient. I will write about this later in the book (see chapter 13), but suffice it to say that while the staff at Emory Healthcare had the heart and courage required, the unit was far from prepared. I sent an email to Dr. David Stephens discussing my concerns. He was receptive and—as Patty secured resources for Emory Healthcare while I set up the isolation unit—secured gear from the closed training center and trained the health care providers. After the training was complete, I was truly honored to be asked to stay with the unit while they treated Kent Brantly and Nancy Writebol, the first two Ebola patients ever treated in the United States.

When the Ebola virus was no longer detected in the blood of Kent and Nancy, I resigned from Emory University. Later, as I was consulting in Belgium and learning from Médecins Sans Frontières, I watched on television the nurses and doctors I had worked with say goodbye to Kent Brantly. It brought tears to my eyes and remains a highpoint of my life.

I continue to consult and provide biosafety training to many around the world. In fact, I have opened a small training center of my own based on the training center at Emory. Here, I will continue to train small groups of professionals for as long as I can.

One of the lessons my mentors Lee Alderman and Henry Mathews taught me serves as the backbone of all I do, not only in biosafety but in everything I do professionally. We serve, with you and by you, increasing your ability to practice safer behaviors for yourself, increasing your capacity rather than a dependency—today and in the years to come. Let's do safety together. I am here to serve, and if you want to discuss anything in this book, please call me directly at 404.849.3966.

Sean G. Kaufman

chapter 1

Infectious Disease Pioneers

Pi.o.neer (noun)
One that begins or helps develop something new and prepares a
way for others to follow: one of the first to settle in a
territory : an early settler

—Merriam-Webster Unabridged
(http://unabridged.merriam-webster.com)

In this book, you will meet Beth Griffin, a young woman working with non-human primates at Yerkes National Primate Research Center in Atlanta, Georgia; Linda Reese, described as the "lab mom," verifying a dangerous sample at the Michigan Department of Health Laboratory; and Maybelline (last name unknown), a maternal delivery nurse in Liberia during the 2014 Ebola outbreak, praying as she struggles to live. These women were infectious disease pioneers who died doing their jobs.

You will also learn about Henry Mathews, a scientist at the Centers for Disease Control and Prevention (CDC), who became sick with hepatitis B while working on samples from Africa during the early days of the HIV crisis; Joshua Gurtler, a United States Department of Agriculture (USDA) scientist, who became sick with *Escherichia coli* O157:H7 while working to make food safer for the public; and Nina Pham, a Texas nurse treating an Ebola patient, who also became sick with Ebola. These individuals, now living with conditions that impact the rest of their lives, are infectious disease pioneers, too.

If you work in a clinical laboratory, research laboratory, ambulance, emergency room, intensive care unit, or isolation room, then you, too, are among the ranks of the infectious disease pioneers. You are on the frontline of emerging and reemerging infections, working directly with infectious pathogens and infected patients for the common good.

The pathogens are invisible. There are newly emerging and reemerging pathogens that are changing and mutating constantly. Although we understand a great deal about pathogens and pathogenesis, there is even more that we do not know. Infectious disease pioneers are not only some of the first to encounter an unknown pathogen, they (you) may also have the earliest exposures and be among the first patients.

I think that this kind of work is heroic. Many don't think of laboratory workers when the word "hero" is mentioned. Typically, that word conjures images of firefighters running toward flames to save lives, police officers providing security of mind and body, and members of the military standing guard. The value and heroism of these individuals are correctly recognized by society for the way in which they put their lives on the line for others. But what happens when these heroes get sick?

Imagine this scenario: One of these heroes wakes up at 2:00 a.m. with a high fever. She is greeted at the emergency room by a nurse who screens her, asks her questions, and guides her to a room. A doctor then visits her and orders blood to be drawn for testing. A clinical microbiologist, either in the hospital or at a nearby lab, receives the sample (along with thousands of other patient samples) to perform the ordered tests, and guides the nurses and doctors in their patient care. More than 24 hours later, everyone involved with the patient learns that she is infected with the severe acute respiratory syndrome virus (SARS). All these nurses, doctors, and laboratorians have unknowingly encountered a very dangerous pathogen just by doing their jobs. They are infectious disease pioneers.

There are more heroes connected to this ecosystem. How do the clinical laboratorians know what tests to perform, how to interpret them, and what recommendations should be made to the health care staff? Identifying solutions to infectious diseases is the important role of research scientists working in research laboratories. They spend hours designing, performing, repeating, and analyzing experiments and encountering risks for the common good. Whether researching the elements of specific pathogens that threaten human, animal, and plant health, studying existing preventive measures and treatment protocols, or searching for new vaccines and medications, a research scientist is also an infectious disease pioneer.

Public health laboratory staff are infectious disease pioneers as well. As soon as health care workers and epidemiologists become aware of an outbreak, public health laboratory staff work to identify the culprit. Public health laboratories identify and distinguish community threats, determine whether they are natural or a form of bioterrorism, and monitor for emerging as well as existing public health threats. Public health laboratory staff are responsible for detecting and alerting other agencies and the public when something is wrong and when increased attention and assistance are needed.

I wrote this book to serve primarily those working in clinical, research, and public health laboratories, because these are the people I train globally. However, as I will discuss later, my time at Emory University Healthcare's

Ebola isolation unit gave me a unique perspective. I saw firsthand how beneficial the application of the biosafety principles and practices used in clinical and research labs could be for the treatment of sick patients by clinical care providers. With more than 1 million health care-associated illnesses and more than 100,000 deaths from these in the United States annually (https://www.ncbi.nlm.nih.gov/pmc/articles/PMC6245375/), it is clear why clinical care providers must add another component to their infection control practices.

Infectious disease pioneers from all of these arenas work together to improve and protect the lives of people around the world. Each contribution, whether in the clinic or in the clinical, research, or public health laboratory, is needed to complete the puzzle of global health. You may not find yourself on the front page of the newspapers (unless you make a mistake), but you certainly deserve to be there!

This book provides guidance for the everyday laboratorian as well as suggestions and guidelines for those in leadership and administrative roles. Most chapters feature a personal story from an infectious disease pioneer or someone close to them. Some stories are explorations or reflections of various career paths that infectious disease workers have taken, whereas others serve as cautionary tales that highlight the very serious risks these pioneers may face. In addition to providing education on safety considerations for those on the frontlines of the infectious disease battle, the book introduces some of the infectious disease pioneers I have had the privilege to know and work with over the years.

 ## BIOSAFETY *in the First Person*

The Loss of Beth Griffin by Caryl Griffin

> *When I first heard Beth's story, I was a new parent. My kids are now teenagers, getting ready to leave the house to start their own adult lives. I cannot imagine the pain Beth's family went through when she was lost. However, Beth's story did not end with her life. As far as I am concerned, she remains the heart of biosafety. Caryl continues to give her heart by telling Beth's story in the hope of protecting people and ensuring that Beth's tragedy never happens again.*

In 1997, Beth Griffin was an artistic, intelligent, and compassionate young woman who worked as a research assistant at the Yerkes Primate Research Center studying hormonal influences on macaque behavior. Her exposure occurred while working within an outdoor compound containing 100 rhesus macaques. Beth was performing physicals on each of the monkeys during an annual roundup. During this process, the monkeys were brought out six at a time in a transport cage and then transferred one at a time to a squeeze cage where they were anesthetized to be examined safely.

The last monkey to be examined that day was frightened and refused to enter the squeeze cage. As Beth leaned toward the cage, attempting to coax the frightened monkey, the monkey jumped and flicked material with his tail from the bottom of the cage into Beth's unprotected eye. It was not common practice at the time to wear goggles.

She removed brown material from her eye with a damp paper towel, and when she asked about washing her eye (she had not yet been trained in eye-wash station use, and this was also not common practice at the time), she was instructed to continue the examination of the monkey because "We all get splashes in the eye." Forty-five minutes later, she rinsed her eye under the faucet in the ladies room with the help of the housekeeper. She was told that no incident report was necessary.

From the time of exposure, Beth was worried about the possibility of contracting the herpes B virus, which is endemic in macaques. Ten days later, on a Saturday, matter began oozing out of her exposed eye. Over the next six days, Beth made persistent, repeated attempts to get help. She first went to her internist, who referred Beth to the emergency room. There, she was diagnosed with pinkeye (conjunctivitis) and told that there was no need for testing for the herpes B virus.

On Monday, the Primate Center's occupational health nurse required Beth to complete an incident report that stated, "No follow-up necessary." That day, Beth called the hospital's infectious disease department directly, asking for evaluation, but was told, "A physician must make the referral." She again called her internist, who referred her to ophthalmology instead of infectious disease. The ophthalmologist diagnosed cat scratch fever and treated her with doxycycline, without testing for herpes B virus.

The next day her symptoms worsened, with a pounding headache, and for the fourth time she called her internist, who referred her to the doctor on call, who referred her to the ophthalmologist the next day. Overnight, she developed photosensitivity, a shooting, pounding headache, and nausea and vomiting.

On her follow-up visit, the ophthalmologist called the infectious disease physician, who immediately admitted Beth to the hospital, where herpes B virus testing was performed and treatment with intravenous acyclovir was begun. Three days later, a central line was inserted, and she was placed on ganciclovir as well as total parenteral nutrition.

Two weeks later, her symptoms improved, and she was discharged with a central line. But overnight, her neck pain, which had begun two days before discharge, increased, and by morning she was unable to move her legs. She was rushed to the emergency room and admitted to the intensive care unit with ascending disseminated myelitis. By early evening, she was intubated, and by morning she was paralyzed to the level of her C2 vertebra. It was thought by her physicians that the herpes B virus infection was resolved. She was treated for postviral autoimmune disease, was given massive doses of intravenous steroids, and received five days of plasmapheresis. On full life support, she died two weeks later.

Since Beth's death, our family has worked to prevent similar tragedies while supporting responsible scientific research. We established a nonprofit organization, the Elizabeth R. Griffin Foundation, in her name to continue this effort. In 2018, the Foundation became the Elizabeth R. Griffin Program at Georgetown University's Center for Global Health Science and Security, building on the Foundation's legacy promoting safety in research and clinical laboratories worldwide.

chapter 2

Understanding Containment

Containment in both laboratory and clinical settings comes down to how the environment is engineered to handle infectious disease threats. In biological laboratories, there are four main biosafety levels. Each level is designed and engineered to house specific infectious disease threats. Clinical containment is a similar but more theoretical practice of managing the infectious disease threats in health care settings.

BIOLOGICAL LABORATORY CONTAINMENT

Biosafety level 1 (BSL1) laboratories are designed to house infectious agents that typically don't make healthy humans sick. Keep in mind, however, that people living with compromised immune systems are more vulnerable and need to take adequate precautions when working in BSL1 laboratories.

BSL2 laboratories are designed to house infectious agents that are spread via blood, fecal, and/or oral routes of transmission. It is important to point out that the greatest threats in public health from an infectious disease aspect reside in the BSL2 laboratory. Scientists typically work with the agents of malaria and influenza and HIV in BSL2 laboratories. However, there are exceptions, and I address those below.

BSL3 laboratories are designed to house infectious agents that are spread via aerosol routes of transmission, such as the agent of tuberculosis. Tuberculosis is an enormous public health threat and may be the most common laboratory-associated infection in the world. In BSL3 labs, directional airflow (from clean to dirty) and biosafety cabinets are available. These two engineering controls offer protection to both the laboratory staff and outside environment. However, many laboratory staff, primarily in underresourced countries, work with tuberculosis outside biosafety cabinets and in BSL2 laboratories. The lack of resources prevents both access to and maintenance of equipment designed to ensure containment of biological agents.

BSL4 laboratories are designed to offer ultimate protection to laboratory staff, the environment, and the general public. Multiple redundancies are

Table 2.1 Biosafety levels

Level	Appropriate use	Example agents
BSL1	Undergraduate and secondary educational training and teaching laboratories and other laboratories in which work is done with defined and characterized strains of viable microorganisms not known to consistently cause disease in healthy adult humans	*Bacillus subtilis, Nigeria gruberi,* infectious canine hepatitis virus, and exempt organisms under the NIH guidelines
BSL2	Clinical, diagnostic, teaching, and other laboratories in which work is done with the broad spectrum of indigenous moderate-risk agents that are present in the community and associated with human disease of varying severity; with good microbiological techniques, these agents can be used safely in activities conducted on the open bench, provided the potential for producing splashes or aerosols is low	Hepatitis B virus, HIV, *Salmonella,* and *Toxoplasma*
BSL3	Clinical, diagnostic, teaching, research, and production facilities in which work is done with indigenous or exotic agents that have a potential for respiratory transmission and that may cause serious and potentially lethal infection	*Mycobacterium tuberculosis,* St. Louis encephalitis virus, and *Coxiella burnetii*
BSL4	Work with dangerous and exotic agents that pose a high individual risk of life-threatening disease that may be transmitted via the aerosol route and for which there is no available vaccine or therapy; agents with a close or identical antigenic relationship to BSL4 agents also should be handled at this level, and when sufficient data are obtained, work with these agents may continue at this level or at a lower level	Viruses such as Marburg virus and Congo-Crimean hemorrhagic fever virus

Adapted from Chosewood LC, Wilson DE (ed), *Biosafety in Microbiological and Biomedical Laboratories*, 5th ed,. U.S. Department of Health and Human Services, Public Health Service, Centers for Disease Control and Prevention, National Institutes of Health, Washington, DC, 2009.

in place, as these laboratories contain life-threatening infectious agents for which there are limited treatment options. It is important to understand that BSL4 laboratories are built not so much for the threats we know about today but for the threats that could be coming tomorrow. We must have laboratories that can provide complete protection to those working to learn about the infectious disease threats we face.

A summary of the different biosafety levels is found in Table 2.1.

Now, let's put your knowledge of containment to the test. If you are working with Ebola virus, what biosafety level containment would you need? If you said BSL4, you are incorrect. The answer is, "It depends." The decision depends on many different factors. Are you dealing with a live strain? Are you processing a suspected patient's sample? Are you working with only a component of the virus and not the entire virus?

Here's a second challenge. Where do you need to be to work with HIV? If you said BSL2, you are wrong. The answer to this question is also, "It depends." What are you doing with the HIV? Are you working with large quantities? Is it very virulent? You need to consider whether you need to increase safety by placing your work in a BSL3 laboratory. Biosafety levels are chosen not just according to the agent but also on the basis of multiple dependent factors.

There are also BSL2+ laboratories, which are basically BSL2 laboratories with staff behaving as though they are working in a BSL3 lab. They will usually implement standard operating procedures (SOPs) like working in biosafety cabinets or wearing additional personal protective equipment (PPE) that you would normally see in a BSL3 laboratory.

Last but not least are the animal facilities using different animal biosafety levels (ABSL) and large-animal facilities, which are usually referred to as biosafety level Ag. These facilities are designed to mitigate risks associated with infectious diseases and animals, large and small. Containment is not just about ensuring that laboratory staff remain safe while working; it is also about making sure that the outside world remains safe as well. Remember, we live in one world, an ecosystem where plants, animals, and humans interact. Although we tend to focus on human health, without healthy plants and animals we humans will die. This is why ABSL and Ag facilities rely on strategies to ensure containment for both human and animal health.

CLINICAL CONTAINMENT

I can talk about containment in biological laboratories very easily. This practice is documented in the literature, and I have seen it in action, although my summary above does not address the complexities. However, *clinical* containment is a practice that I am proposing. It is a theory—a vision—that I hope will be incorporated into infection control strategies in health care facilities.

Biosafety offers much that can be utilized in infection control. One of the greatest benefits could be the engineering of the environment where sick patients and health care providers interact. But designing the environment is just the beginning; consideration of how infection control could be implemented using SOPs, PPE, and leadership controls is desperately needed.

A classic example can be seen in the common emergency room. People come into an emergency room for a variety of reasons. However, those who need stitches and X rays are treated via the same processes as those who feel sick. Those who feel sick may have a fever and yet sit in the same area and chairs as those who are not sick. Furthermore, most emergency rooms have vending machines, inviting those who may not be sick to eat after touching surfaces that could have been contaminated by those who are sick.

Infection control processes are effective and bring tremendous substance to health care facilities. However, we can do much better. I believe that this starts with a proper design of facilities and triage of patients who are exhibiting symptoms of an infectious disease. A lesson will continue to present itself until it has been learned. The deaths and illnesses among health care workers responding to Ebola and Middle East respiratory syndrome, two emerging infectious diseases, should be a strong wake-up call. Here some brief descriptions of clinical containment levels.

Clinical containment level 1 (CCL1) is the general hospital setting that requires health care providers and patients to practice good hand sanitation. These settings don't require additional engineering, because healthy people usually would not get sick from coming into contact with the microbes generally found there. However, people who are fighting cancer, have received an organ transplant, or are living with a compromised immune system can get sick from common agents found in the CCL1 setting. Patients with a compromised immune system should be provided with extra precautions and separated from other ill patients when visiting a health care facility.

CCL2 is engineered to protect providers against infectious agents that are spread via blood, fecal, and/or oral routes of transmission. Health care facilities should provide not only proper PPE and SOPs but also an environment with secured access that prevents those who don't belong in the area from accidentally entering it.

CCL3 is engineered to protect providers against infectious agents that are spread through aerosol routes of transmission. Directional airflow, high-efficiency particulate air (HEPA)-filtered exhaust, single-pass air, hands-free sinks, double-door entry, sealed windows, and a sealable environment should be provided when treating patients with these conditions. Again, engineering controls alone would not suffice, because health care providers must have access to PPE and follow strict guidelines when working within this containment level. Additionally, health care facilities must screen and triage patients quickly to prevent contaminating environments or exposing health care staff unnecessarily.

CCL4 is engineered to protect providers against infectious agents that are life-threatening and for which there are limited treatment options. All practices used in CCL2 and CCL3 facilities should be implemented in addition to waste management processes, HEPA-filtered air supply, liquid waste management processes, interlocking door systems, and a sealed environment. This environment should also include personal locker rooms, an anteroom, and a decontamination room, allowing health care providers the opportunity to properly decontaminate on the way out of the area where the patient is being treated.

Clinical containment levels are summarized in Table 2.2.

Although I am not a health care provider and cannot tell anyone the best way to implement these strategies, I know that they are needed, for many reasons. First, they formalize a process that, based on risk, allows health care providers to utilize engineered spaces for the treatment of infectious disease. Second, once an environment becomes risk specific, protocols that include PPE, waste management, and patient treatment can be

Table 2.2 Clinical containment levels

Level	Appropriate use	Example agents
CCL1	Patients who may be ill with microorganisms not known to consistently cause disease in healthy adult humans	Nonpathogenic *Escherichia coli*, other noninfectious bacteria such as *Bacillus subtilis*
CCL2	Patients who may be ill with indigenous moderate-risk agents that are present in the community and associated with human disease of varying severity	Hepatitis B virus, HIV, *Salmonella*, *Clostridioides difficile*, methicillin-resistant *Staphylococcus aureus*, influenza virus
CCL3	Patients who may be ill with indigenous or exotic agents with a potential for respiratory transmission that may cause serious and potentially lethal infection	*Mycobacterium tuberculosis*, highly pathogenic influenza virus, severe acute respiratory syndrome virus, Middle East respiratory syndrome virus
CCL4	Patients who may be ill with dangerous and exotic agents that pose a high individual risk of life-threatening disease for which there is no available vaccine or therapy	Congo-Crimean hemorrhagic fever virus, Ebola virus

developed to complement an engineered space. Finally, if we engineer spaces and design strategies to contain clinically acquired infections, we can begin training health care providers to prepare for the next emerging infectious disease.

Health care facilities must be able to identify infectious patients more quickly than they do now. They must take adequate precautions for themselves and others while the infection is being assessed. Once assessment is complete, health care providers know what they are facing and must have a place where they can send the patient to be managed. This management must offer protection to the provider as well as the patient.

This dream of blending laboratory and clinical containment can be realized if we check what we think we know at the door and ask ourselves, "How can we do this better together?"

BIOSAFETY *in the First Person*

My First Patient by Dottie Cappola-Vojak

Dottie is an amazing nurse. She has served during many emergency response situations and today survives both the emotional and physical effects of responding during the events of September 11, 2001. In her story, Dottie demonstrates how those working on the frontlines of infectious disease rarely place their health and safety above the service they provide to others. This human factor—a call to serve—must be recognized, as it poses one of the greatest risks in safety today. Dottie reminds us with her story that the fight is not just with an infectious disease but with human spirit as well. Dottie is a fine example of an infectious disease pioneer hero.

I did not get the chance to go to nursing school until I was 35 years old. I had three young boys, worked as a lab tech, and always dreamed of being a nurse. With the help of God, and with much fear, I enrolled in the nursing school attached to the hospital where I worked. They had a program for employees that paid for one class per semester. I jumped on it, and three years later, in 1988, I graduated, passed my boards, and was a registered nurse. I stayed at this hospital but now as a nurse. Being a "new nurse," I, of course, was assigned the night shift. I worked 11 p.m. to 7 a.m. five days a week. It was difficult, but on the plus side, I was finally a nurse, and working five nights in a row gave me the opportunity to really get to know my patients.

My very first patient was Jean, a 23-year-old man from Haiti. Jean had come to this country when he was only 10 years old. He was smart, polite, and handsome and had a smile that would light up a room. Jean told me all about Haiti and his big family there, who were poor and lived in a small village where there was no school. He told me how fortunate he was to be able to come to America with his mother. They lived with his aunt and uncle, who had come from Haiti years before and had opened their home to him and his mom. He went to school for the first time at 10 years old but he quickly caught up with the rest of the class and went on to graduate from high school. He got a job at a local food bank and spent his days trying to help others less fortunate than himself. He'd laugh at that, saying who would ever think anyone could have been less fortunate than him when he was in Haiti? But just look at what he had now!

Jean's heart and mind were so full of life and hope. But Jean was very ill. He had end-stage HIV disease. Very little was known about HIV at this time, and there were few treatment options. We basically treated his symptoms. Jean was in "contact isolation," which I had learned in nursing school meant that I had to wear gloves and a paper gown. I followed the rules the best I could. His body was shutting down, including his kidneys, from the effects of the HIV. The fluids in his body were not being filtered out properly, so they were pooling, first in his feet, then his entire legs. Soon, he was hard as a rock from the waist down from built-up fluids. It was very painful for him. Jean suffered from severe nausea, vomiting, and diarrhea. He was always in pain. But he always greeted me, every shift, with that beautiful, winning smile. He was grateful for whatever I did for him.

I heard from the day nurse that Jean rarely had visitors. He had now been in the hospital for two months. He was alone. I tried to spend as much time as possible with Jean, but I had six other patients to care for each shift. When Jean would call out in pain, or if he had had profuse diarrhea, I would run into his room as soon as I could. Often, I was too late, and he had completely soiled himself and the bed. Trying to get to him quickly, I would sometimes not stop for a gown but just throw on a pair of gloves. When I finished cleaning him up, I ripped off my gloves, and whatever was on them would sometimes splash onto me, my face. But this was nursing. This was what I did. There was never any fear of catching HIV. I was safely covered most of the time.

Jean's disease soon consumed him. He was on a morphine drip for the pain, and I had to continuously increase the dose to try to keep up with that pain. I was with Jean the night he died. I knew the moment I entered his room that night. There was no smile. Instead a look of fear and pain was on his face. I went to him. His morphine was at the maximum dose. I could not increase it for him. He was cool to the touch, yet his face was covered with sweat. I got a cool cloth, wiped his face and his neck. He looked at me. His eyes were distant, not focused at first. Then he recognized me, tried his best to smile. He knew, and I knew, that the end was near. I adjusted his position and put my arm around his shoulders. He let his head rest on my shoulder. I just talked to him. Talked about the stories of his life in Haiti that he had told me. About how much I enjoyed those stories. I also told him about my family, about being the oldest nurse in my class in nursing school, and that how unfortunate he was that he was my very first patient and that I really didn't know what I was doing. How I prayed every day that I wouldn't make any mistakes and that I was doing the very best for him. Jean weakly told me that he was blessed and that God got me to him "just in time."

Jean died in my arms that night. As I prepared his body, he was not an HIV patient or a hepatitis patient. He was Jean, my first patient and now my friend. I reflected on what he had gone through, what we had gone through, over those months. The world was full of fear of HIV. Many patients were suffering more from the way they were treated than from the disease itself. As nurses, we learn about infectious disease. We learn about symptoms, treatments, contamination, and "isolation." And we always try to do the right things the right way. But, as every nurse knows, they are our patients. We are here to care for them the best we can. We have to leave the fear at the door and replace it with warmth, love, and compassion. As I wrapped Jean's body, I admit I shed some tears. I hoped I had done my best for him. I said good-bye and then walked away to care for another patient, thanking God for the skill, knowledge, and heart to be a nurse.

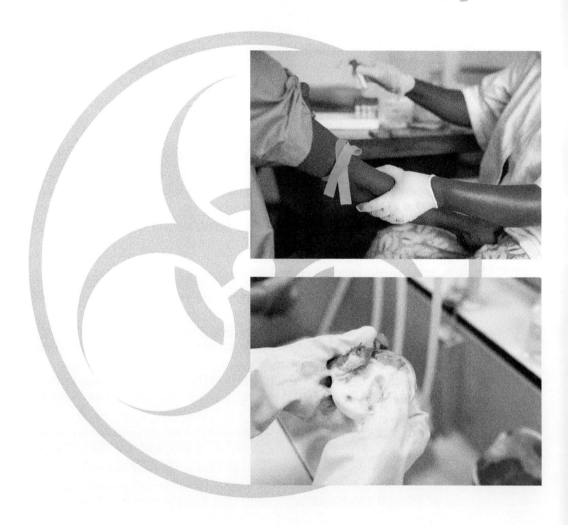

chapter 3

Biological Risk Mitigation

We will never be able to eliminate risk completely. However, the fact that you are reading this right now means that you have been successful in reducing all kinds of risk in your life. You are a natural risk mitigator.

PHASES OF BIOLOGICAL RISK MITIGATION

There is no secret to biological risk mitigation, but there is a useful tool for remembering the four phases of the process: "I AM Communication," standing for "identify, assess, manage, and communicate." These four phases can be used to mitigate risk not only in a laboratory but in our lives in general.

Frequently, when I hold a training session, regardless of where we are in the world or what language the person is speaking, I have the following conversation:

Q: What do you do before you cross the road?
A: I look both ways.

Q: What are you looking for?
A: Cars or anything else that could hit me!

Q: If you see a car, what do you do?
A: I wait.

Q: What are you waiting for?
A: The car to pass.

Q: Yes, I see. But the car is parked. Why are you waiting?
A (usually with a smile): Of course!

We have become so inherently good at risk mitigation that, once a specific hazard has been identified, we typically don't even consciously think about the risk assessment process. Malcolm Gladwell, in his book *Blink*, explains that humans can process large amounts of information about specific risks within a very short time (usually instantly) and come to an accurate conclusion about the overall risk. Therefore, we need to be aware that one of the biggest human risk factors (which we will cover in detail in chapter 4) is the fact that our minds do not naturally think in a scientific fashion. What could take years to discover and conclude using the scientific method of experimentation, the human brain does amazingly fast!

Going back to the dialogue that I frequently have during training:

Q: After you have done your assessment of the street conditions, you have three choices. What do you think those choices are?

A: I can stay where I am. I can walk across the street. I can run across the street.

Q: What if you have children or friends with you? Do you communicate what you are doing?

A: Usually, especially if I am with children.

Q: So you think it is important to tell children why you are doing this?

A: Of course! It teaches them how to do it for themselves.

Again, no matter where I am in the world, the results of this discussion are almost always the same. We as humans naturally mitigate risks. Because of our natural inclination to do this, we may not always pay attention to where the risks are, and we can improve our overall risk mitigation strategy, as I will discuss in the following sections.

EVOLUTION OF BIOLOGICAL RISK MITIGATION

I like to think we mature collectively as a society. If this is true, I also propose that we mature collectively in a profession as well. This is certainly what I have observed during my time serving the profession of biosafety. I have witnessed the growth of the profession, led by expert teachers who in turn have shared stories of their own observations on how the biosafety profession has matured over time. The next section will briefly cover this maturation. As mentioned, the four phases of biological risk mitigation are hazard identification, risk assessment, risk management, and risk communication.

Hazard Identification Era

In the early days of the biosafety profession, the focus was on the most obvious hazard of all: the biological agent that could hurt us. Once the agent was identified, a scientific assessment would determine appropriate behaviors or management strategies to allow scientists to work safely with and around the agent. The thinking was that the more we got to know the agent, the safer we were. Unfortunately, knowledge alone does not equal increased safety.

Although the agent itself is the most obvious hazard, it is not enough to consider the dangers of the agent in isolation. My first 10 years working in biosafety were mainly focused on instilling in the workforce the belief that, if they were prepared and had knowledge about the agent they were working with, they would be safe. However, this strategy turned out to be not enough.

Think about it. The agent is really on vacation when it comes to the laboratory. Very rarely do we see a scientist attempting to agitate the agent. It is as though it has come to a resort. When we want it to grow, we feed it and keep it warm. When we want to put it to sleep we place it in a freezer. When it is time to retire, we have an autoclave for the big send-off! It is an easy life for the agent. Going beyond identifying the agent as a hazard is very important.

We began to understand that the way the agent is worked with could pose a great risk, so a number of questions began to present themselves. Though the agent lives a stable life, we as humans do not. Sometimes we have good days—we are engaged and well rested. Other times we may hate what we do or who we work with, or we are tired. These inconsistencies affect how we do what we do. Furthermore, we need to explore what procedures are being used by individuals working with the agent. What equipment and resources do they have at their disposal? Are the workers trained? Are they overconfident and complacent in their work? The answers to these questions opened the doors to identifying additional risk factors that had not been previously addressed.

We now understand that even the most educated and prepared workforce can and will practice unsafe behaviors if the safety culture of the organization is weak. There are many examples of competent professionals behaving poorly—not because of insubordination, apathy, or complacency, but rather because of the social norms of the community they work in. In other words, if the culture they work in is one that lacks expectations and accountability with regard to practicing safety, then their behavior will correspond with that, regardless of their background or training.

In summary, hazard identification at a minimum must include assessments of the agent, the people working with the agent, and the culture of the organization. Many other hazards can be identified, but these three are the most important ones.

Risk Assessment Era

After the hazard identification era, there came a phase in which all problems seemed to be solved by doing a risk assessment. It was as if the motto were "In Risk Assessment We Trust." Risk assessment is certainly the backbone of biosafety and should be included in all activities when people are working with biological agents. However, there are some gross oversights in risk assessment that need to be addressed.

I quote my colleague Ren Salerno, who pointed out that risk assessment is not something you do once a year, it is something you do every single time you enter the laboratory. It is a lifestyle for those working around biological agents. Today, risk assessment has become a formal process, appearing to many as a chore rather than a necessity. This perception must change; risk assessment is a critical tool for identifying necessary equipment, developing protocols, and verifying workforce preparedness levels both prior to and during work with biological agents.

Another issue to remember regarding risk assessment is that we ourselves constitute a risk factor. Regardless of the results of a risk assessment, please be aware that (i) we as humans determine our own levels of personal risk and (ii) failure to assess our perceptions of the risk will produce an incomplete picture of the overall risk. In September 2006, I was saddened by the death of Steve Irwin ("the Crocodile Hunter"). I had enjoyed watching him wrestle crocodiles, handle the most poisonous snakes, and do things that made me think he had a death wish. However, he was an expert in what he did; unfortunately, this time he encountered a stingray. Now, to him the stingray was not something to be too concerned with. However, to the general population, snakes, crocodiles, and stingrays are a risk. Rather than running toward these animals, many would run away. This shows that an individual's personal perception of his or her risk impacts his or her safety attitudes and behaviors around the risk. It also shows that we may have vast differences in how we perceive the same risk.

The final factor to consider about risk assessment is the value to the end user. A risk assessment attempts to identify the probability and severity of a specific hazard. There are several formulas and theories that can yield such data. On completion of the risk assessment tool of your choice, do you know what to do with this information? The risk assessment data alone, without the understanding of how to use them, are essentially worthless. Risk assessment is, and always will be, a vital part of biosafety, but it is useful only if the results of the risk assessment are understood and applied to the goal of managing risks in safer ways.

Risk Management Era

The natural successor to the risk assessment era was the risk management era. Risk management focuses on taking the results of a risk assessment and using them to create detailed and meticulous standard operating procedures (SOPs) to address the safety issues identified in the risk assessment. When I ask a safety training audience, "How many of you have SOPs for writing SOPs?" I always get chuckles. As the profession of biosafety matures, organizations are using risk assessment findings to develop strategic plans for mitigating risks. This is a great start, but strategic plans alone are not going to get us to safer behaviors.

An SOP is truly an SOB (standard operating behavior). Unfortunately, many organizations seem to know more about writing plans than they do about those who will be asked to follow the plans. Organizations can forget the human factor; they expect an individual to follow the plan regardless of the level of experience or education he or she may have. They forget that, even if experience and education are verified, variables within the person and the environment will produce different attitudes and behaviors that affect the safety of both the individual and those working around that individual. Writing plans and presenting them to the workforce constitute only a portion of what is needed for safety. Because we expect people with different backgrounds, experiences, and education levels to behave consistently, we must provide them with what they need to attain this behavior. This leads us to the final era of biological risk management, the communication era.

Risk Communication Era

Humans are efficient creatures. We generally don't behave a certain way or do something unless there is a need for it. Despite this, most organizations today develop plans and place behavioral expectations on the workforce without providing reasons for them. We are asking people to do something and not explaining why they need to do it.

During safety trainings, I have met many people who could not explain why they should not remove their hands from a biosafety cabinet after touching a contaminated item, why a biosafety cabinet needs to be certified, why eyewash stations and eye protection are needed, or why handwashing and not eating in the laboratory are important. You may think that avoiding these behavioral mistakes is just common sense, and I agree. However, I return to the caution about being human. Unless the requested safety behavior provides a benefit that is obvious to the person performing the behavior, he or she will simply ignore the request and perform his or her main task as efficiently as possible.

Here is another aspect of communication for you to consider. During the Leadership Institute for Biosafety Professionals, I asked biosafety professionals to identify their high point of the year with regard to biosafety. Dee Zimmerman, a biosafety officer for the University of Texas Medical Branch (UTMB), offered this event: losing vials of one of the most dangerous agents in the world. I couldn't believe my ears—how was that a high point? She explained it this way: "The discovery of the loss was not an indicator of failure; it was an indicator that our biosafety program was working. For one, the scientist at the center of the loss trusted us, informed us, and engaged with us during the investigation." Leadership at UTMB went public with the news, making headlines around the world. They defended their scientists and their program and were open, honest, and transparent about everything. Today, UTMB remains a trusted member of the community because they communicated quickly and frequently and showed an understanding of how the public would react to such an unexpected event.

The profession of biosafety will inevitably shift to focus more on better communication with both the workforce and public. As mentioned, this begins with explaining why we have certain behavioral expectations of those to whom we are providing SOPs and extends to informing the public about unexpected occurrences. The public can tolerate risk, even when it comes to working with biological agents, but when something unexpected happens specific to that risk and the event is hidden, suppressed, or otherwise covered up, the public becomes angry and distrustful. Healthy communication about why we do what we do and about unexpected happenings ensures a stronger relationship and more trust for the future.

The four phases of biological risk mitigation are never-ending. As hazards and risks shift and change, we must continually identify, assess, manage, and communicate. Looking back on the earlier eras of the biosafety profession, we can see how each phase taught us important lessons and has brought us to our present level of safety maturation. Just like the eras of biological risk mitigation, the profession of biosafety will continually adapt, change, and evolve.

BIOSAFETY *in the First Person*

The Lab Mom by Carrie Anglewicz

> *Carrie tells a very special story, one that she related during a training program I was facilitating. The loss she experienced so many years ago still remains with her today. Carrie's story brings Linda's memory alive, especially for all of those known today as "lab moms." I continue to share her story with many around the world.*

Linda Reese passed away from laboratory-acquired meningococcemia on December 25, 2000. Linda had worked at the Bureau of Laboratories in Michigan for 28 years. She was a skilled laboratorian, a friend and mentor, and, above all else, a wonderful person.

I started my career at the Bureau of Laboratories in 1998. It was my first real job out of college, and I had dreamed of working there since I toured the facility while in school. My duties included working in two different sections, and I spent three days each week working in the Enterics/STD/Chromatography (ESC) unit. I appreciated the variety of work, and my coworkers were smart, fun, and friendly. I had no local friends when I started, so having good colleagues meant a lot. One of these colleagues was Linda Reese.

Linda's primary duty at the lab was *Salmonella* serotyping, a very specialized job. She did this work for 15 years and truly was an expert. At one time, the Bureau of Laboratories in Michigan had produced its own antiserum for *Salmonella* testing, and Linda knew how to absorb it into a working form. This is a process that is a dying, if not dead, art. The *Salmonella* bench is a busy one, but Linda made time to get to know me and made sure I was settling in. I remember my first Christmas working there: Linda and another colleague bought me a star for my Christmas tree because I didn't have one. That's just how she was.

By 2000, my duties had changed, and I was working solely in the ESC unit on the *Neisseria* bench, which was close to Linda's bench. We received a fair amount of nongonococcal *Neisseria*, including *Neisseria meningitidis*, which I would serotype if the sample was from a sterile site. I would joke that I was the "queen of gonorrhoea." I like to make people laugh, so I would make up songs or tell stories just to hear Linda chuckle. I can still hear her laugh and say with incredulity, "Okay, Carrie," at something ridiculous I had said.

I came to consider Linda my "lab mom." She had two teenage daughters of her own but had many times invited me to her home (she had the fattest cat I'd ever seen!) and to events outside work. Linda was very active in her church, and she once brought me with her to a social function there. She even volunteered each week at the local elementary school helping children with reading.

Some people have an unhappy resting facial expression; they may look mad or sad when concentrating or doing nothing. Linda was the opposite; she had a "resting happy face." Even when she was quietly working on her *Salmonella* samples there would be a little grin on her face. It may sound like Linda was a perfect angel, and that's because to me she was.

Linda's birthday was December 23. She passed away two days after her 52nd birthday. At the time, our unit rotated who would work on holidays. State offices would close for an extended time, but we all know that bacteria don't take holidays, so someone would need to tend to them. Holiday shift work included logging in new specimens and doing any necessary testing, including *Neisseria* identification and serotyping. All other specimens would be stored frozen and revived when normal work hours resumed. It was Linda's turn to cover the Christmas holiday in 2000, and with the way the holiday fell, her scheduled work day was December 22. That year we all chipped in to reserve a room at a local restaurant for an after-work Christmas party. I think it was December 21. That was the last time I saw Linda alive.

My family is not local, and I take vacation between Christmas and New Year's Day. I have done this every year since 1998 except for one year, 2000. I reported for work on December 26 at 8:00 a.m. Linda was scheduled to start her shift at 7:00 a.m., but she was not in when I arrived. This was unusual for two reasons. First, she was never late without calling. Second, it was her week to rotate on a different bench, which she liked to do as soon as possible, so she could leave on time.

Around 8:30 a.m. the phone rang, which was not unusual since we had been closed for several days. Clinics and sentinel labs were anxious for their results. I still have the visual memory of sitting at my computer and looking at the phone as it rang. I got up to answer it, fully expecting to hear from a hospital laboratory.

I don't remember exactly what Mike, Linda's husband, said when he told me she had died, but I do remember asking him to repeat himself. I truly didn't think I had heard him correctly. Looking back, I am still sorry for how painful it must have been for him to have to say it twice. Mike gave me an account of what had happened. Linda had prepared Christmas Eve dinner for her family but had gone to bed early because she didn't feel well. It was later reported that she had a 104°F fever. She woke up on Christmas morning, but her condition deteriorated rapidly. She was taken to the hospital and passed away shortly after her arrival. As Mike relayed the information to me that her illness was thought to be meningitis, part of my brain disconnected from him and panicked, filled with questions, "Am I getting this right? What do I do? What happened to Linda?"

Our supervisor, Bill, arrived while I was on the phone. He had just opened the door to his office, which was near the phone I was using. I'm certain I said my condolences to Mike and assured him someone else would be in touch. I knew some of the management hadn't started their day yet, and they'd be the ones to ask the real questions. I felt like I wasn't the one who should have taken the call. What if I got information wrong and told everyone something untrue? I simply didn't believe what I had heard from Mike.

And how would I tell our supervisor? Bill was a great boss, a soft-hearted man. I was worried about how he would take the news, so I did the only thing I could think of. "Bill, sit down," I said. Did I think he would faint or that his knees would give out, like you see on TV? I don't know, but it seemed like the right thing to do. He gave me a funny look and I said it again, "Sit down." Then I had to tell him. Being young and in shock myself I had no clue how to be gentle, so I just said it. When I finished, Bill walked into his office and closed the door. My mind was still reeling. I must have looked visibly shaken because I remember people asking me what had happened, and I had to tell them, too.

I can only imagine what was happening in the closed-door meetings of management. Over the next couple of weeks, there were inspections and media attention. There were investigations by us, the State of Michigan, the Occupational Safety and Health Administration, and the CDC. It took some time for the details to become clear. Linda had died from *N. meningitidis* serogroup C. Pulsed-field gel electrophoresis was used to link it to a specimen that she had worked on during her holiday shift. Linda was a conscientious scientist. A colleague who was also in the lab on December 22 said that Linda hadn't reported any incident or unusual occurrence. The truth is that we will never know what happened to her and how she was exposed. At the time, our laboratory tested samples for *N. meningitidis* at biosafety level 2 containment, which was standard practice. All of the investigations found no lapse in our practices. I can't tell you exactly how many *N. meningitidis* samples I had serotyped in the same way Linda had, but it was a lot.

Practices for testing suspected *N. meningitidis* have changed since Linda's death, not only in our laboratory, but across the United States. Vaccination is now standard for anyone working with it, which was not true in 2000. Would it have saved Linda? Maybe, maybe not. Was she immunocompromised at the time, putting her at greater risk? It's possible. Regardless, I'm glad we have made those changes, both in our lab and nationally. Linda's husband, Mike, is quoted in the *Lansing State Journal* as saying, "I am sort of gratified. It means she didn't die entirely in vain."

Mike's quote is part of why I tell Linda's story to biosafety professionals. I also tell it to give a personal account of what happens when things go wrong. A serious laboratory-acquired infection is not a single event. It affects everyone in the lab. First and foremost, Linda was our friend and an asset to the community. Following her death, I took on Linda's *Salmonella* serotyping duties. What a challenge it was to learn this skill, this art, without her expertise. We learned a lesson in that regard, too: make sure that your experts share their skills. Cross-training lab members is key. Although I had to learn without her, I came across Linda's handwriting on vials of antisera for years and on documents I kept for reference, bittersweet mementos.

My colleagues and I were haunted by the "what ifs." "What if the specimen had come in a day earlier—could I have prevented this, or would I have died instead?" This is another component of laboratory-acquired infections, one that can't be reported in *MMWR* or other scientific journals. One might call it collateral damage. The lab member who becomes sick or injured isn't the only one whose life is affected. Despite this darkness, my colleagues and I came together to support each other, and we were also supported by others. I was reminded that, despite the tragedy, this was a great place to work and the work that we do is important.

Linda Reese will always be remembered by her friends, family, and colleagues. At the Bureau of Laboratories, she is remembered in lasting ways through a memorial garden and a keynote presentation in her name every year during national lab week.

chapter 4

Human Risk Factors

Human risk factors are conditions or behaviors arising from our natural qualities as humans that put us at risk for certain adverse outcomes. Understanding human risk factors can assist in reducing organizational costs associated with incidents, accidents, and missed deadlines. The more we understand how humans behave in an environment within the context of the culture to which they belong, the more we can be proactive in our overall safety strategies. Addressing human risk factors requires the application of multiple fields of learning, including psychology, engineering, statistics, and anthropometry (the study of measurements and proportions of the human body).

One of the biggest human risk factors may well be the fact that generally we don't understand much about ourselves, our capabilities, and our inadequacies. As you read this chapter, remember that no matter who you are, where you live, or what language you speak, if you are a human, these risk factors apply directly to you, to your behavior, and to those around you. It is human behavior that connects strategic biological risk mitigation plans—the standard operating procedures (SOPs)—with desired safety outcomes. Until robots replace humans, it is a good idea to understand the humans working with and around infectious disease agents.

THE SWISS CHEESE EFFECT

What do you see when slices of Swiss cheese are stacked one on another? Usually, the holes do not line up, preventing you from seeing the plate through the multiple slices. However, occasionally by chance, the holes line up perfectly. We can think of human risk factors in a similar way, with each cheese slice representing a different layer of biosafety, and each hole representing an individual risk factor. When the different human risk factor "holes" align, there are no barriers to catastrophe. When two holes line up, it is usually no big deal, but when multiple human risk factors match—watch out!

In this chapter, you will meet Jasper, an imaginary laboratory staff member with whom we will explore human risk factor concepts. Jasper works in a research laboratory with biological agents that can be spread through fecal, oral, and blood routes of transmission. As we examine Jasper and his behaviors, imagine the slices of Swiss cheese stacking up, with the holes lining up perfectly. Remember, because you are human, you could be Jasper.

APPEALING TO AUTHORITY

The social psychologist Stanley Milgram demonstrated something very troubling about humans during his obedience experiments in the 1960s. He concluded that humans have the ability to dissociate their behaviors from the impact of those behaviors on others when they are ordered (by an authority figure) to do something, even if the behaviors are unsafe or malevolent. This finding should always be included in any discussion about human risk factors.

Jasper is a doctoral student at a major university, and his principal investigator (PI) has asked him to do something he knows is restricted research. Jasper is not comfortable with the request but is told that if he chooses not to do what is requested, then he will not receive his doctoral degree. Jasper jumps into the research, dissociating himself from the work he is doing and placing all blame on the PI who is making this demand of him.

This ability to separate oneself from responsibility for one's behavior is a primary human risk factor and poses challenges to both biosafety and biosecurity. One control for this human risk factor would be to put into place an anonymous reporting system that allows anyone within the institution to report any demand he or she considers inappropriate. All claims should be investigated, and those who report them (whistleblowers) not only should be protected but should receive positive reinforcement for reporting.

Jasper, for the sake of his degree and research career, chooses to move forward with the restricted experiment. His organization does not have a mechanism for reporting his discomfort over the research request, and there are very few controls in place to regulate authority figures. Let's see what happens next.

ENVIRONMENTAL CONDITIONS

Although Jasper likes to believe that he is in control of his environment, the environment can control him as well. By environment, I mean the culture and atmosphere created in an individual's working situation. Jasper's institution sets clear expectations for its workers but does not hold laboratory staff accountable for failing to meet these expectations. In other words, when SOPs are not followed and, instead, laboratory staff behave based on their own perceptions of risk, the situation goes unnoticed and nothing is said or done to address this. With variation of behavior around risk comes variation of outcomes. Jasper sees that his institution does not respond when others in his environment do not follow SOPs, and so he begins to behave the same way. He uses his own experiences and education to steer his behavior, rather than following a specific risk mitigation strategy, i.e., the SOPs.

Jasper's organization also chooses to punish staff who report incidents, accidents, or near misses. The science is conclusive as to the effects of punishment on behavior. Punishment aims to stop behavior but must be applied in a clear, immediate, and consistent fashion for it to work. But what does "work" mean?

We know that punishment does three things. First, it builds resentment. If we assume that a person behaves in a certain way because he or she has some need that is being met by this behavior, then by asking them to stop this behavior, we are taking away their ability to meet that perceived need. Abraham Maslow's hierarchy of needs observes all human behavior as responses to challenges in meeting our basic needs, and I agree with this observation. To explore this concept, imagine that a parent tells his child to stop picking her nose in public. The child looks at the parent with a sense of resentment and wonders why he would make that request. The child is picking her nose because she has a nasal obstruction and is having trouble breathing comfortably. The parent is literally telling his child to stop breathing. So, she waits for an opportunity to sneak away, and once out of sight of the parent, she begins picking her nose again.

This is the second outcome of punishment—it hides true behavior. When you punish people without addressing what is driving the behavior, they only stop doing the behavior when you can witness it. Because the need that caused the behavior in the first place is still present, they will continue the behavior when you are not there to punish them. They behave when you are looking, because their need not to be punished trumps the original need, but when the punisher is absent, they return to the behavior.

The third outcome of punishment is that it addresses only the negative behavior—*stop picking your nose!*—and it doesn't teach a replacement or new, acceptable behavior. If we believe, per Maslow, that behavior is driven by a challenge in meeting a basic need, then the way to address poor behavior is to provide alternative behaviors that meet the basic need.

Humans are not separate from their environment and the associated culture, they are a part of it. If an organization has expectations for its workers but does not hold them accountable for their behavior and punishes those who report issues, the workers will hide safety issues that could cause significant risk to themselves and others.

Jasper's organization does not hold people accountable, and when something goes wrong, the laboratory staff are punished, via embarrassment, firing them, tarnishing their reputation, and/or stopping them from doing their research. How likely would Jasper (or any of us) be to report an incident or accident? The answer is, of course, that we are not at all likely to speak up until something catastrophic occurs.

MENTAL STATE

Within the laboratory environment, what is Jasper's current state of mind? As a human, Jasper learns things through a process called the stages of learning, an adaptation by Cole, Harris, and Field of the Prochaska stages-of-change model. Jasper starts out with a mental state of thinking he knows more and can do more than he actually can. As you can imagine, if Jasper was learning how to swim, this could be very dangerous. Jasper might jump into the pool thinking he knows how to swim, only to learn as he is drowning that he cannot. It is just as dangerous if a laboratory staff member learns that he or she is lacking a necessary skill while in the middle of working with a biological agent. The human risk factor associated with this mental state can be controlled through skill audits, which are assessments that ensure that laboratory staff can competently perform the tasks that are expected of them.

Another mental state that contributes to the human risk factor is belief that laboratory staff members have been trained and that is enough. The word "trained" should be removed from our lexicon, because nobody is ever fully trained. The fact is that we are constantly training, by being present and thoughtful in our daily work activities. Acknowledging this puts us in a healthier mental state and demonstrates an appreciation for continued lifetime learning and behavioral evolution.

While I was growing up, my father frequently told me, "Mess up, and mess up often. If it doesn't kill you, it will make you better." This is a common cliché, but it expertly summarizes the concept of behavioral evolution, the cycle through which we constantly acquire and improve new skills. The first stage of behavioral evolution is learning a new behavior. Do you remember the first time you rode a bike, drove a car, or entered a laboratory environment? The first time we do anything, we tend to pay attention to all of the details. Our neurological system is fully engaged; we lack confidence and are intentionally slow in our actions and thoughts.

However, after a short period, we enter the second stage of behavioral evolution, called perceived mastery. This stage is a nightmare for parents who are teaching their children how to drive. It is the point at which radios begin to blast and children begin texting as they drive. They believe they have mastered the behavior of driving—until something goes wrong. The same thing happens in the laboratory when someone who has been "trained" begins to feel confident at this stage. They develop habits and patterns of behavior until something unexpected occurs: an accident, incident, or near miss.

The range of consequences of the perceived mastery stage is quite variable, from a near miss in which nobody gets hurt, all the way to an incident in which serious injury or death occurs. Fortunately, as humans, we have the ability to learn vicariously. This is partially why I chose to include stories of such incidents as part of this book. It is my hope that the consequences of the perceived mastery stage, either from a personal incident or one you read about in a story, change you and increase your self-awareness over time. Despite increased awareness, with time, complacency usually sets in again and habitual practices take control—until another potential near miss, incident, or accident occurs. This cycle is one way that humans naturally change their behavior over time.

So how do we combat this cycle of perceived mastery and complacency? One option is to break up routines, change daily responsibilities, and include training exercises and scenarios that challenge daily patterns and habitual behaviors. This is behavior-based training. It attacks the mental states of perceived mastery and complacency before near misses, incidents, and accidents have the chance to occur.

Jasper's institution, which lacks accountability and utilizes punishments when incidents and accidents are reported, has an official training policy. Employees are initially trained when they first join the lab, and follow-up training occurs annually through a computer-based system. Jasper received his initial training and now is in a stage of perceived mastery regarding his restricted experiment. As a result, he has had several near misses, but for fear of punishment has not reported them. The Swiss cheese is stacking up.

BIOLOGICAL AGENT–HUMAN INTERFACE

Science has the ability to quantify potential risks, but most people determine their own individual risk based not on the science but instead on their personal history and experiences. This, in and of itself, is a human risk factor because the small sample size of our personal experiences around a risk can make it appear less risky than it actually is. As mentioned, when we first work around a new risk, we tend to over-respect the risk. During this time, we evaluate our risk based on how it has been quantified by science. However, as time passes and our comfort increases, we become over-confident, believing that the personal risk is lower than that quantified by science. This type of thinking explains variances such as those ranging from new scientists who want to wear three pairs of gloves when working with Ebola virus to those who believe they may not even need gloves. There is a wide range of perceptions specific to biological agents. These perceptions are created directly by our experiences.

When I was young, my father cut the seat belts out of his cars because they were ugly. Years ago, dentists and doctors used to forgo protective gloves when performing procedures, from filling cavities to performing Caesarian sections. Laboratory staff used to pipette by mouth and eat in the laboratory environment. These risks were taken based on personal experiences that led to false perceptions of risk.

This major human risk factor can be controlled for with the use of SOPs. SOPs are not just for the new staff but rather for all staff, regardless of experience and education. SOPs are not to be used some of the time; they are meant to be used every single time work is being done around a biological agent. SOPs control for human risk factor variances in education, experience, and perception.

As you can imagine, this control depends not only on how good an SOP is but whether the SOP is followed, going back to the idea of accountability that we discussed earlier. A later chapter in this book is dedicated to the SOP because it is a critical safety component in all laboratory environments. However, putting in place an SOP that increases overall risk or fails to mitigate a risk is not just wrong, it is unethical. Make sure your SOPs work!

Jasper is in a stage of perceived mastery with a restricted experiment and has not reported the several near misses that he has experienced. He has been given SOPs, but he has been working around this particular biological agent for years without any negative incidents or experiences, and so his perceived risk with the agent is quite low. As a result, Jasper starts taking shortcuts in his safety procedures.

PHYSICAL STATE

Our bodies are not static. Factors such as age, stress levels, and both acute (short-term) and chronic (long-term) conditions can have a direct influence on our immune systems and their ability to fight potential exposures to the biological agents with which we are working.

Many organizations require facility maintenance staff to undergo physical examinations to identify preexisting conditions before hiring. Yet many laboratory staff are hired and begin working with infectious biological agents without being screened for preexisting conditions that could affect their ability to fight the infections they could get while working in the laboratory. As our health is not always the same from day to day, it would make sense to screen staff daily or weekly for fevers as well as screening them annually for new medical conditions that develop. Not doing these health checks is a serious shortcoming of laboratory safety in general.

We must also consider the medications laboratory staff are taking on a chronic basis or begin taking while working in the laboratory. Some medications inhibit the immune response, making a possible exposure and infection worse than they would be for someone with a healthy immune system. Also, consider the people who are living with the laboratory staff, such as elderly parents with compromised immune systems, a partner fighting cancer, or children with autoimmune diseases. Even pets could be affected by what is being worked with in the laboratory.

Jasper is working with a restricted experiment, not reporting incidents, and taking shortcuts. He does not seem to respect the risk involved with his work. He has recently contracted a bronchial infection that is being treated with a steroid. Each of these individual uncontrolled risk factors alone might not lead to a negative outcome, but coming together as they are, like the pieces of Swiss cheese with holes aligning, they presage a devastating outcome for Jasper.

INDIVIDUAL CAPABILITY

Individuals have different capabilities. Depending on what the job expects of the laboratory staff, some individuals may not be able to cognitively or behaviorally live up to those expectations because they lack the capability to do so.

An SOP alone cannot produce consistent behavioral outcomes among different people with different levels of education and experience. Only when we focus on the behaviors and capabilities needed to follow the plan do we achieve the outcomes we are looking for. For example, if we require laboratory staff to wear specific personal protective equipment (PPE), they may need to have a certain level of physical ability to don or doff it safely. Many individuals do not have enough flexibility or balance to remove PPE without falling or hurting themselves. Others may have discomfort or phobias that render them unable to wear specific pieces of PPE.

I believe in making sure that scientists demonstrate safe practices at biosafety level 2 (BSL2) before entering BSL3, and likewise that they demonstrate good practices at BSL3 before entering BSL4. The reason for this is that higher levels of containment require more attention to detail and compliance with SOPs and the ability to respond strategically to unexpected situations. This typically requires a great amount of experience gained through working, training, and mentorship. It also requires demonstration of a level of cognitive capability.

Just because someone wants to work with biological agents does not mean that he or she should. Desire cannot be the only prerequisite for gaining access to a biological laboratory. Similarly, just because someone is capable today does not mean that he or she will stay that way. Eyesight can fail, and dementia, distraction, or other mental or physical impairment can develop that affects our ability to safely do our job. Laboratory staff must be able to demonstrate the physical capability to work safely. They must also demonstrate attentiveness, discernment, and fortitude—the cognitive capabilities needed to work safely in a laboratory environment.

As we continue to stack the Swiss cheese slices of human risk factors, Jasper's situation is getting worse. He is taking shortcuts while working on a restricted project and not reporting incidents that are leading to exposures. Jasper is sick and taking a medication that compromises his immune system, and because of his constant shortness of breath he is struggling to wear his PPE properly. Jasper begins choosing when to wear PPE, based not on the SOP but on his own capabilities and lowered perceptions of risk associated with the agent with which he is working.

EMOTIONAL STATE

Emotions affect our overall state of being and have a direct impact on our perceptions of a situation. Emotions typically are split into two categories: pleasant and unpleasant. Within the pleasant and unpleasant categories there are active and inactive states to which we should pay close attention.

Just as for the physical state, there are acute and chronic emotional states. Depression is an example of a chronic condition in the unpleasant category that affects the emotional state of an individual. At the other end of the spectrum, an acute change in the emotional state of an individual in the pleasant category could be caused by having a baby or getting married. Although the categories are at opposite ends of the spectrum, they each could have detrimental effects on the worker, causing distraction or new detrimental behaviors.

Because of Jasper's physical state, his emotional state is one of fatigue. His Swiss cheese slice risk factor holes have now all lined up. His carelessness with his PPE leads inevitably to an exposure to the restricted agent. Since his immune system is compromised, this exposure results in illness. At first, Jasper does not report his condition because of the culture of punishment that his organization fosters. His condition worsens, and ultimately Jasper succumbs. Now, finally, the entire stack of Swiss cheese is revealed. The public outside his organization learns that a restricted experiment was going on there and that organizational failings have resulted in a young man's death, and people respond with anger, fear, and mistrust. This casts a dark cloud over all of the science going on at Jasper's organization, and the organization itself becomes at risk for being shut down.

Although each individual risk, considered in isolation, seemed fairly innocuous, you can see how easily they compounded into a catastrophic outcome. To prevent personal and professional tragedies, we must implement controls for each of these human risk factors. Performing science safely requires not only understanding the biological agents being worked with but also knowing our own behaviors, perceptions, biases, and limitations.

 BIOSAFETY *in the First Person*

Malcolm Was My Friend by Joseph Kanabrocki

Joseph Kanabrocki is the Associate Vice President for Research Safety and Professor of Microbiology in the Biological Sciences Division of the University of Chicago. Joe is humble, reflective, and extremely kind—qualities you would want both in a friend and in a safety professional. Joe is committed to the profession of biosafety and to protecting future generations of scientists. The story that he relates about his friend and colleague Malcolm Casadaban demonstrates how multiple failures of what seem to be irrelevant and unconnected factors can indeed line up perfectly, producing a worst-case scenario.

My career as a microbiologist began in graduate school, where I worked in a P2 laboratory, the old equivalent of BSL2 in more contemporary vernacular. My dissertation examined elements associated with the symbiotic relationship between the protozoan *Paramecium tetraurelia* and the bacterium *Pseudomonas taeniospiralis*, neither of which is pathogenic to humans. My postdoctoral studies continued in the molecular biology of *Paramecium* and other nonpathogenic microorganisms, *Tetrahymena* and the yeast *Saccharomyces*. During the course of my graduate and postdoctoral work, my biosafety training can be summed up in two words: sterile technique. Sterile technique is essentially biosafety turned inside out, preventing any potential contamination or threat from the outside environment from entering the research area or reaching the sample being worked on. We were taught to painstakingly adhere to sterile technique to protect the quality of our research, while not thinking much about safety. This is not terribly surprising given the lack of biohazardous microorganisms in our research program.

So how did someone who worked with harmless microorganisms get into a career in biosafety? It was during my second postdoctoral appointment that I was introduced to biosafety. A colleague and dear friend, John Bade, the Executive Administrator of the Laboratory of Molecular Biology at University of Wisconsin at Madison (UW–Madison), showed me a job posting in the local newspaper. The UW–Madison biosafety officer (BSO) was retiring, and recruitment to hire his replacement was under way. It took me a while to stop laughing at John's suggestion that I apply for this position. I didn't want to be someone who earns a living issuing citations! No, thank you! John persisted over the course of the next several months, pointing to my ability to deal with people as being the most essential element of the job, and revisited it on nearly a weekly basis when the new classified ads were published in the Sunday paper. So, apply I did. Much to my surprise, I was granted an opportunity to interview with 13 UW faculty and staff over the course of two days. During the interview process, I began to see what a career in biosafety might look like and realized that I really wanted the job.

Roughly two weeks later, UW called and offered me the job as BSO. As it turned out, I was taking over the lead role in a very mature biosafety program. My predecessor had served as UW–Madison BSO for 20 years, with his assistant BSO, Betty Sullivan, serving with him for 19 of those years. The best news of all was that Betty was not retiring. Although I was technically Betty's boss, I was smart enough to realize that Betty was my mentor and that I was the mentee. For the first several years of my career, Betty ran the show and I learned from her. I told myself that if ever I could find a source of funding, I would start a biosafety fellowship program modeled after the mentoring I received at UW–Madison. This dream became a reality many years later at Washington University in St. Louis (WUSTL). At the time, WUSTL was the home institution for the National Institutes of Health (NIH) National Institute of Allergy and Infectious Diseases (NIAID) Midwest Regional Center of Excellence for Biodefense and Emerging Infectious Diseases Research. Later, I moved to the University of Chicago and was able to clone WUSTL's biosafety fellowship program. The University of Chicago was also the home of another institution, the Great Lakes Center of Excellence for Biodefense and Emerging Infectious Diseases Research. Over the 10 years of NIAID funding, we were successful in training 10 postdoctoral fellows, all of whom have gone on to either lead biosafety programs or consult on biosafety programs for other institutions. It is unfortunate that the program ended; we need more programs like this in the biosafety community.

My first external biosafety program audit happened approximately three months into my new job as a BSO at UW–Madison. One of our investigators was funded by the Department of Defense (DOD), and the DOD was sending their lead BSO, Robert Hawley, to Madison to review the lab's program. Bob turned what could have been a very difficult and contentious inspection (this particular lab was not in very good shape) into a teaching moment, not only for me, but for the principal investigator and all of the investigators and students in this lab.

Any biosafety officer worth his salt is a member of the American Biological Safety Association (ABSA). It was through ABSA that I was able to meet with and learn from the biosafety giants who came before me: Emmett Barkley, Byron Tepper, Manny Barbeito, and the above-mentioned Bob Hawley, to name but a few. All of these great men were scientists first, and many worked in the U.S. Biodefense Programs at Fort Detrick. (I am told that the term "biosafety officer" was born at the U.S. Army Medical Research Institute of Infectious Diseases, as those responsible for biosafety often met at the officers' club to discuss their work.) They all recognized the value and importance to society of life sciences basic research, especially in the realm of infectious diseases, and the role of the BSO in facilitating this science. Each of these biosafety pioneers emphasized, and in fact mandated, evidence-based approaches to biosafety.

Just as all BSOs know ABSA, we all also know the CDC/NIH publication *Biosafety in Microbiological and Biomedical Research Laboratories*, more commonly referred to as the BMBL. Together with the *NIH Guidelines for Research Involving Recombinant or Synthetic Nucleic Acids*, the BMBL provides a framework for self-governance of life sciences research by scientists. This framework for self-governance dictates that scientific inquiry must be conducted responsibly and ethically, traits that are more vital today than ever before due to the rapid technical advances not only in the life sciences but also in disciplines such as synthetic biology that have emerged by exploring the intersection of engineering, computing, and life sciences.

Among other things, the BMBL discusses laboratory-acquired infections (LAIs). The primary objectives of any biosafety program are to mitigate the risks associated with investigations involving pathogenic microorganisms, to minimize to the greatest extent possible exposures to these pathogens, and to prevent LAIs. But no amount of study or diligence had prepared me for an LAI that happened on my watch as BSO, much less a fatal one.

Malcolm Casadaban, Associate Professor of Molecular Genetics and Cell Biology and Microbiology, was my friend and colleague at the University of Chicago. His last day in the lab was Monday, September 7, 2009. He phoned labmates telling them that he was ill and that he would not be coming in to work. His illness progressed, and he was seen in urgent care on Thursday, September 10, where he presented with flu-like symptoms. He never mentioned to his health care providers his work with *Yersinia pestis*, a CDC-designated select agent and the causative agent of plague. Malcolm most likely didn't mention this to his doctor because his work involved exclusively an attenuated, non-select-agent *Y. pestis* strain called Kim D27. This strain was deemed to be so attenuated, so safe, that at one point it had been explored for use as a plague vaccine. Malcolm was sent home by the doctor and told to rest and drink plenty of fluids. On Sunday, September 13, Malcolm's condition continued to worsen, and he was taken by ambulance to the University of Chicago Emergency Department. His respirations became more and more labored and he required intubation. Malcolm passed away soon thereafter.

I was on travel the week of September 13 and was told of Malcolm's passing upon my return on Tuesday, September 15. The next day I attended Malcolm's memorial service at the University of Chicago Bond Chapel, with many of Malcolm's students, colleagues, friends, and family in attendance. Malcolm was 60 years old and an insulin-dependent diabetic at the time of his death; I attended the service assuming his death was due to natural causes. However, on the day of his memorial service, blood cultures taken in the emergency department revealed a septicemia, with early characterizations indicating the presence of Gram-negative rods and Gram-positive cocci. Preliminary characterization of the Gram-negative clinical isolate was *Yersinia pseudotuberculosis*. This was our first clue to the possibility of an LAI.

On Friday, September 18, I was summoned to a meeting in the dean's office. It is not often a good thing when you are summoned to an unannounced meeting in the dean's office late on a Friday afternoon. There must have been at least 20 people in this meeting, the vast majority of whom I had never met or even seen before, including clinicians, lawyers, and media and communications staff. It was in this meeting that we first discussed the finding that Malcolm's death was due to an LAI.

But how could this be? Malcolm worked with *Y. pestis* Kim D27, a "harmless" strain. Did he somehow get his hands on the fully virulent, select agent *Y. pestis* strain called CO92? How could this have happened? Malcolm did not have clearance to work with virulent *Y. pestis* CO92, nor did he have access to the secured select agent BSL3 lab that housed the CO92 strain. Was this a breach of security? Or instead, was the virulent CO92 released from BSL3 containment? Was this release an accident, or could it have been intentional? Was this a biosafety failure, an accident, or a malicious event? Were others who worked with or near Malcolm at risk? What about the clinicians who worked to save Malcolm prior to his passing? Were we witnessing the unfolding of a lab-derived outbreak of the plague in the United States? As I sat in this meeting, numb to what I was hearing, all of these questions were racing in my head.

Immediately following that meeting, a feverish, around-the-clock effort to characterize the clinical isolate began and continued through the weekend. Another meeting followed, early on Monday morning. In this meeting were more than 40 people, including all of the same attendees as the Friday afternoon meeting as well as representatives from state and local health departments, Illinois and Chicago Departments of Health, the CDC (including the Select Agent Program), federal and local law enforcement, the FBI and the Chicago Police Department, and a representative of the Occupational Safety and Health Administration (OSHA). Everyone around the table was looking to me for answers as to what happened; I was the BSO, after all. Among the many items discussed in this meeting was a debate between the Chicago Police Department and the Chicago Department of Public Health (CDPH) as to who would lead the investigation, with both agencies arguing why they should take the lead. This debate continued until the meeting group decided to visit the lab in which Malcolm worked. The lab had been closed and vigorously surface-decontaminated over the course of the weekend. Upon seeing the biohazard signage on the lab entrance, the Chicago Police Department decided that maybe the CDPH should take the lead in the investigation after all.

To say the scene was chaotic is an understatement. There were more than 40 visitors, all looking very uncomfortable in their issued lab coats, walking through a research lab, most not knowing what to look for. Upon assessing the situation, the OSHA investigator pulled me aside. His first words to me were, "My condolences on your loss." This was the first person to show empathy about the loss of our friend and colleague. He then went on to say, "You obviously have your hands full. I will reach out to you in about two weeks after the dust has settled a bit." Among all of the authorities who had descended upon the University of Chicago for this investigation, only the OSHA representative seemed to understand what we were experiencing.

Those who had come into contact with Malcolm prior to his death were offered antibiotic prophylaxis. Autopsy revealed an enlarged liver with iron deposits, the first clue that Malcolm might have had hemochromatosis, a genetic disorder that manifests itself as an iron imbalance, usually as a person gets older, and resulting in accumulation of iron in organs and tissues. Postmortem blood chemistries confirmed highly elevated iron amounts in Malcolm's blood and tissues, and genetic testing later confirmed that Malcolm indeed had suffered from hemochromatosis, about which he was not aware. The relevance of this finding is that the elevated iron levels in his body enabled *Y. pestis* Kim D27, which derives its avirulent status from being defective in its ability to acquire iron from its environment, to replicate in Malcolm as if it were the fully virulent *Y. pestis* CO92.

Characterization of the clinical isolate, including eventual sequencing of its genome, confirmed that the clinical isolate was indeed the exact *Y. pestis* Kim D27 strain that Malcolm had engineered himself. While we now knew that there was no breach of security, no inadvertent release of a pathogen from BSL3 containment, we did not know (and to this day still do not know) how

Malcolm became infected. Interviews with labmates suggested that Malcolm's biosafety practices were lax, with this old-school microbiologist often seen picking colonies from petri plates with toothpicks while wearing neither gloves nor lab coat. When asked, "Why did you not say anything? Why was this behavior not reported?" the labmates, many of whom were grad students and postdocs, said, "Who are we to question a tenured faculty member?"

This tragic event taught me a number of important lessons. First, no matter what your occupation, your primary care physician should be aware of your work and, most importantly, what hazards you face during the course of your workday. In particular, if your work involves potential exposures to disease-causing agents, the possibility of a work-related illness should always be considered. Second, it is vital for each of us to be aware of our own health status, and we must recognize that this is a moving target, particularly as we grow older. It is vital that we recognize that when our health is compromised, whether transiently due to a community-acquired infection or more permanently as a result of a physiological deficit, it is a very poor decision to put oneself at even greater risk by doing work that requires rigorous attention to detail. Third, it is everyone's responsibility to watch out for each other and to report accidents and incidents that deviate from established SOPs, and to ensure that these incidents or near misses result in punitive action only if they are *not* reported or are a result of gross negligence. These accidents and near misses are opportunities to learn, and it is vital that these lessons be shared so that others might learn from them before potentially experiencing the event firsthand.

chapter 5

Four Primary Controls of Safety

Most people spend their lives avoiding infectious diseases, using vitamins, vaccines, medications, face masks, and other strategies to stay healthy. But infectious disease pioneers are eager and willing to do their jobs even when their lives are at risk. The public may believe they are crazy for wanting to work with and around infectious diseases, thinking of these pioneers as "adrenaline junkies" or as taking absurd risks with large drums of biological agents. The truth of what infectious disease pioneers do is further clouded by portrayals in Hollywood movies such as *Outbreak*, in which the handling of biological agents is shown in ways that are completely unfounded in reality. Unlike those rogue infectious disease workers portrayed in films, the day-to-day work of most real-world infectious disease pioneers revolves around precaution and careful planning. The four primary controls of safety, which I discuss in this chapter, provide a framework of multiple safety redundancies for infectious disease pioneers working in biological laboratories or health care facilities to minimize risks to acceptable levels.

A definition of the word "safety" can be found in any dictionary, but how do we turn a definition into something that can be applied? Three things must be blended to form a complete safety program: (i) safety guidelines, (ii) the four primary controls of safety, and (iii) the three levels of applied safety.

GUIDELINES

Safety guidelines identify what the acceptable level of safety is within any given laboratory or health care facility. The CDC and National Institutes of Health (NIH) are the agencies in the United States that issue biosafety guidelines in the form of the publication *Biosafety in Microbiological and Biomedical Laboratories* (BMBL). These standards provide the biosafety guidelines for biological laboratories. However, the BMBL is not the only set of biosafety guidelines. The World Health Organization provides biosafety guidelines that are more appropriate for under-resourced countries. Several other countries have also written their own appropriate safety guidelines, with each country's unique issues addressed. For example, biosafety level 3 (BSL3) laboratories in the United States must have single-pass air. This means that air comes into the laboratory in a one-directional flow and is immediately exhausted out, not recirculated within the laboratory. However, the current WHO biosafety guideline states that BSL3 air can be recirculated within the laboratory if it is first filtered using a high-efficiency particulate air (HEPA) filter. These differences in guidelines reflect differences in climate and resources; for many areas, air conditioning is unavailable due to cost and therefore single-pass air is simply not practical.

PRIMARY CONTROLS

Safety guidelines provide the direction needed specific to the four primary controls, which are (i) engineering, (ii) personal protective equipment (PPE), (iii) standard operating procedures (SOPs), and (iv) administrative or leadership controls (Table 5.1). When implemented properly, these four primary controls provide multiple redundancies for laboratory and health care facility staff working with infectious substances.

For example, the BMBL states that all work in a BSL3 laboratory must be done in a biosafety cabinet. This guideline is chosen because most of the agents worked with in BSL3 laboratories are spread through aerosol routes. The biosafety cabinet itself is an engineering control, the first of the four primary controls. If the engineering control fails—for example, if there is a spill outside the biosafety cabinet—the second control provides additional protection. The individual working in the biological safety cabinet wears PPE, which protects the portals of entry of a pathogen into the body. If the second primary control is breached, SOPs, the third primary control, provide evidence-based strategies for responding to the spill and doffing potentially contaminated PPE. The final primary control, which actually encompasses the other three, is administrative control. This ensures that staff have been trained to respond to situations, comply with SOPs, have been vaccinated (if available), are monitored for potential laboratory-acquired infections, and receive prophylactic treatment (if available). Together, the four primary controls minimize risk for laboratory and health care facility staff, the reputation of the organization, and the reputation of science overall.

But who determines what minimizing risk means?

LEVELS

This is where the three levels of applied safety come in (Table 5.2). Leaders, safety officials, and laboratory staff all tend to view risk differently. Generally speaking, leaders are concerned with fiscal sustainability and the reputation of the organization. Safety officials are concerned with the health and safety of the clinical and laboratory staff. Laboratory and clinical staff are concerned with their research, level of care, and career progression.

Table 5.1 Four primary controls of safety

Engineering
- Directional airflow
- Hands-free sinks
- Autoclaves
- Self-closing doors
- Biosafety cabinets

PPE
- Gloves
- Eye protection
- Respirators
- Scrubs
- Booties
- Coveralls
- Lab coat

SOPs
- Donning PPE
- Doffing PPE
- Needlesticks
- Spills
- Evacuation
- Waste management

Leadership
- Training
- Occupational health programs
- Medical and incident surveillance programs
- Compliance and accountability programs
- SOP evaluation, validation, and verification
- Vaccination programs

These differences in overall point of view skew how each of these groups perceives risks and, subsequently, lead to differences in how they apply safety.

The three levels of applied safety are (i) acceptable, (ii) preferential, and (iii) ideal. An acceptable level of safety means safety is applied as written in the selected safety guidelines. This requires attention to and semantic differentiation of the words written in the guideline, such as "must," "may," and "should." If something *should* be included, that does not mean it *must* be included. According to the BMBL, a BSL1 laboratory does not need a door, nor does it need screens on windows that open to the outside. The required sink for handwashing and nonporous benchtops and chairs provide an acceptable level of applied safety and engineering controls for BSL1.

A preferential level of safety means safety is applied with one redundancy above what is written in the safety guidelines. For example, the

Table 5.2 Applied safety

Acceptable
- Standard safety
- Minimal costs ($)
- Minimal time invested
- Liberal approach
- No redundancy

Preferential
- Standard safety, +1
- Increased costs ($$)
- Increased time invested
- Prescriptive approach
- Some redundancy

Ideal
- Standard safety, +2 (or more)
- Substantial costs ($$$)
- Substantial time invested
- Restrictive approach
- Multiple levels of redundancy

BMBL does not require work to be done in a biosafety cabinet (BSC) in BSL2 laboratories. If an organization purchases a BSC for a BSL2 laboratory, this is considered a preferential level of applied safety specific to engineering controls.

An ideal level of safety means that safety is applied with two or more redundancies above what is written in the safety guidelines. If the BSL2 laboratory mentioned above elects to install directional airflow along with the BSC, that would be two additional engineering controls above the stated guidelines found in the BMBL, and as such, this BSL2 would be at an ideal level of safety specific to engineering controls.

The higher the level of applied safety, the more expensive and restrictive the laboratory program becomes. Maintenance and sustainability of additional controls are added expenses, and having additional controls also means needing additional SOPs to ensure consistent behavioral practices, which restricts the program in terms of training and time.

Applied safety is the blending of guidelines with the four primary controls, including customization depending on the overall investment preferences of leadership, staff, and safety officials within an organization. No two safety programs are alike; however, all should share the common ground created by the four primary controls.

That was a lot of information to take in, and there's more. The four primary controls are not the sole factor contributing to a successful safety program. Safety programs also require, at a minimum, guidelines and a commitment to an acceptable level of applied safety before the four primary controls can be properly utilized.

ADMINISTRATIVE (OR LEADERSHIP) CONTROLS

The control that has the greatest impact on human behavior and reduction of risk is the administrative, or leadership, control. Included here are vaccination, fit-for-duty screening (both pre-employment and annual), medical and incident surveillance, compliance and accountability, and training programs.

Vaccination

One of the best ways to protect the workforce is to provide vaccinations against the agents with which they are working. However, vaccination programs should not be considered the final line of defense. I have observed instances of organizations investing in vaccination programs for their staff while ignoring the importance of providing proper PPE and appropriate SOPs. There should be no difference in practices or PPE between those who have been vaccinated and those who have not. It is for this reason that I support giving staff the choice as to whether to be vaccinated. If proper PPE and SOPs are being invested in, there is no need to lose a good employee over their choice to not get vaccinated. I am a huge supporter of vaccinations, but reliance solely on vaccinations to protect the workforce is both incomplete and irresponsible. There are many reasons a vaccination may fail. Therefore, we must treat those who have been vaccinated in the exact same way as those who have not been.

Fit-for-Duty Screenings

There are no industry standards that include pre-employment and continuous fit-for-duty screenings for those working with infectious pathogens. As you read in chapters 1 and 4, both Linda Reese and Malcolm Casadaban died from laboratory exposures in BSL2 laboratories. Linda was fighting a respiratory condition and taking steroids as part of her treatment (leading to compromised immune function) when exposed while processing a meningitis sample. Malcolm had an undiagnosed iron overload in his system while working with an iron-deficient attenuated strain of *Yersinia pestis*. Both deaths might have been prevented with effective fit-for-duty screening programs.

This bold statement is meant to be critical of the absence of an industry standard in this regard. Employers have the right to ensure that their reputations, fiscal futures, and workforce members are protected. The type of medical screening depends on a variety of factors. I recommend requiring that any medication or medical condition (chronic or acute) that suppresses immune function or causes seizures, disorientation, drowsiness, and/or confusion be reported. This would not exclude these individuals from being able to work with infectious pathogens, but it would allow leadership to be informed and prepared in the event that a problem occurs.

Medical Surveillance

Medical surveillance and occupational health programs are both important for safety and serve unique purposes, but it important to remember that they are not the same. Occupational health programs provide an overarching service to the entire organization. These programs may include vaccinations, screenings, prophylactic treatments, employee postincident monitoring, and advocacy for patient care. Medical surveillance programs provide a vigilant "neighborhood watch" style program for laboratory staff who develop medical symptoms matching the clinical presentation of agents being worked with inside the facility. Medical surveillance programs capture patterns among laboratory and clinical staff, and if multiple people get sick within the incubation period of the agent being worked with, an investigation for a common source exposure occurs.

Medical surveillance programs don't require substantial resources. All that is required is a committed staff, a contact number, and a log book to report all symptoms the workforce may report. A medical surveillance program within a laboratory or clinical setting can be thought of as being similar to a leak alarm in a chemical or radiological laboratory and is an important part of any safety administrative control program.

Incident Surveillance

Imagine that a spill occurs in the laboratory or health care facility. Two people clean the spill and do not report the incident. This event occurs on a Tuesday. On Saturday, the two individuals present to different hospitals in separate locations. Neither discusses the fact that they work in the laboratory or health care facility, and by the time the organization finds out that two of its employees are in the hospital, one has died and the other is barely healthy enough to report the incident. If an incident surveillance program had captured the spill, the employees could have been monitored in case there was a potential exposure, which helps protect both the employees and the workers at the hospitals to which they were admitted.

All incidents involving infectious substances should be treated as potential exposures, with monitoring of staff until they clear the incubation period of the agent to which they may have been exposed.

Incident surveillance programs also provide a means for organizational improvement, using each incident as a learning opportunity. The difference between a lesson learned and a lesson ignored is change. Incident surveillance programs are the backbone of the continuous improvement process. They allow us to identify what went wrong, understand why it went wrong, and ensure that we make changes to prevent things from going wrong again.

Compliance

An SOP is a behavioral request. Together with the workforce, safety officials and leaders work to develop behavioral strategies aimed at minimizing risks that were identified and assessed as part of the risk mitigation process. If a behavioral request is made, leaders must provide what is needed for laboratory staff to behave in the desired way. Compliance requires both the request to behave and the action of ensuring that laboratory staff have what they need to do so. This is an equal process, one that does not differentiate among laboratory staff regardless of their position, education, and experience. In fact, I strongly encourage leaders to go into the laboratory and intentionally break SOPs. If those in the laboratory witness the leader acting in noncompliance with an SOP and do not address this with the leader, the leader should confront the staff and empower them to do so.

Accountability

As discussed in the previous section, compliance involves setting an expectation. However, compliance alone fails to recognize those doing their jobs correctly, in contrast to those who are not. That is the role of accountability. To discuss accountability, we must discuss the difference between being treated equally and being treated fairly.

I have three children, each of whom has been given a cell phone with the same expectations communicated to them. A straight-A report card is required for them to keep their phones. If two of my children came home with all As (an indicator that they not only worked hard but succeeded in fulfilling my expectation), and one child did not live up to the expectation, what should I do? I should also ask myself, "What are the effects of the actions I choose?"

This is where the terms "equally" and "fairly" are easily differentiated. Our children are capable and were provided with phones and access to computers and the Internet for homework and studying. We monitored their grades, provided feedback almost immediately if grades fell below the A mark, and communicated frequently with all teachers. In other words, our children were treated equally and given the resources necessary to comply with the expectation.

Even though they were treated equally, in this hypothetical situation two children succeeded and one did not. As a leader (parent in the family), it is my job to hold my children accountable, both positively and negatively. Positive accountability means I give something pleasant to those who succeed. Negative accountability means I take something away from those who fail, until they succeed again. The phone was taken away from the child who did not get straight As. I treat the children fairly based on their level of compliance, but that does not mean they are treated equally. If I had treated them all equally (and all three kids kept their phones regardless of their actions), why would the two who succeeded choose to meet our expectations in the future? Leaders must not only request staff compliance with established expectations (guidelines and rules) but must also hold staff accountable, using both positive and negative measures.

A program which ensures staff compliance and accountability, while differentiating between equality and fairness as they pertain to positive and negative accountability, is needed for common behavioral practices to be established among people with diverse backgrounds, educational levels, and experiences. The only way to establish a culture (i.e., a group of individuals behaving similarly and believing in similar things) of safety is to ensure that all members of the organization are compliant with expectations and are held accountable.

Training

The final element of administrative or leadership controls is training. The answers to "Why do we train?" and "Why should we train?" are different. Unfortunately, most organizations train to check a box and get it behind them so they can move along and complete the work that needs to be done. Most people within a workforce exhibit dread, boredom, frustration, and even avoidance when the word "training" is mentioned.

To be honest, I don't blame them. Most training today is delivered through slideshow presentations or online webinars. Unless you have a very engaging trainer, one who refreshes his or her slides frequently and is passionate about the training program he or she is delivering, training can be boring and may also be a complete waste of time. Slideshows and other passive types of training certainly have a purpose. They provide flexibility and automatically generated reports that make it easy for an organization to demonstrate that they've checked certain boxes. They can increase awareness about a specific issue or produce problem-solving opportunities aimed at increasing expertise among staff. However, they do not, and will not, teach anyone the behaviors needed to be safe. Imagine trying to teach someone how to swim using just a computer-based slideshow!

In chapter 13, I provide more specific details and recommendations for improving training programs. But here, let us define training as a program designed to increase awareness, ability, and/or application of safety principles. Training programs should demonstrate that they make a difference in the behavior of those who attend (using pretraining and posttraining evaluations) and should be positively evaluated by those who attend. If the program does not receive positive evaluations, the training program may cause more harm than good. That is, the training program could ultimately make people more resistant to future training opportunities.

Leaders must be the ones to advocate for training, meaning that they must encourage staff to attend and take training opportunities seriously and ensure that they do so. The workforce has a job to do, and when training is planned, it takes time away from the things they must get done. If we assume that an individual works 40 hours a week for an average of

48 weeks a year, that comes to 1,920 total hours. If 5% of this time were spent on training, that would equate to ~96 hours of training, or about 2.5 weeks. However, pulling staff away from their jobs for 2.5 weeks still seems excessive to many leaders. I recommend that 2.5% to 3% of on-the-job hours be spent in training. This comes out to about 48 to 57 hours of training per year, enough time to foster a culture of safety, sustain workforce efficiency, and increase workforce engagement. This is still a considerable amount of time for an individual to be away from his or her daily tasks, and therefore, leadership must take the initiative to schedule trainings, lay out expectations (compliance), and hold the staff accountable to complete the training.

Administrative (or leadership) controls are the main ingredient for building and sustaining a culture of safety within any organization. An organization can invest millions of dollars in engineering controls, thousands of dollars in PPE controls, and hundreds of hours in writing SOPs, but it takes only one instant of noncompliant, untrained behavior to negate all of this.

ENGINEERING CONTROLS

Engineering controls are elements built into the design of a facility, piece of equipment, or protocol in order to minimize risk. Engineering controls may well be the most expensive primary control of safety, not only to purchase but also to maintain. It costs an average of 10% of the total construction costs of a new laboratory or health care facility per year just to keep the facility running. In other words, if you spend one million dollars on a laboratory, expect to pay at least one hundred thousand dollars a year just to keep on the lights on!

When considering the building, design, and device specifications for engineering control purposes, it is important to look beyond the initial costs and consider the overall maintenance and sustainability of what is being acquired. The cost of introducing an engineering control into a design must be sustainable over time. The resulting balanced expense is warranted to ensure that biological risks stay where they are supposed to stay: inside the laboratory or health care facility environment.

Building

Buildings can be designed to include several engineering controls. Examples of these controls include double-door entry, interlocked door systems, HEPA-filtered air supply and exhaust systems, double-door autoclaves, hands-free sinks, self-closing doors, interlocking supply and exhaust fans, Phoenix valves, liquid waste treatment, sealed and sealable rooms, directional airflow, and sealed or screened windows. This is not a comprehensive list of possible building engineering controls but provides the most crucial ones used in most laboratory and health care facility buildings.

Devices

A device is equipment purchased and placed in the laboratory environment. Devices used as part of engineering controls include biosafety cabinets, incubators, freezers, sharps containers, self-capping sharps, shields, small autoclaves, centrifuges, waste containers, and soap dispensers.

PERSONAL PROTECTIVE EQUIPMENT

In 2017, the David J. Sencer Museum at the CDC opened an exhibit that honors the public health officials who served in the 2014 Ebola outbreak. As one of those public health officials, it is an honor to see such a tribute being offered at CDC. However, for me, a dark shadow hangs over this exhibit.

Prior to this this outbreak, I had trained thousands of infectious disease pioneers to work in BSL3 and BSL4 laboratories. My first association with the Ebola outbreak was to train the Emory University Healthcare nurses and doctors who cared for the first two Ebola-infected patients in the United States. I attended and assisted in the training for Doctors Without Borders in Belgium and helped doctors and nurses in Nigeria with Hospitals for Humanity. While I was in Monrovia, Liberia, with Samaritan's Purse, teaching biocontainment strategies, Henry Mathews urged me to review the newly published guidelines from the CDC for donning and doffing PPE by health care providers who are working with Ebola patients. When I read them, I was shocked and disappointed.

The CDC is one of our greatest national treasures, and I along with many of my career mentors have worked at the CDC. As a behaviorist, I know that humans are imperfect beings, and as such, mistakes are made. However, I considered the newly created SOP for doffing PPE irresponsible. In my opinion, the SOP greatly increased risk to those who followed it. There is more about this in chapter 12.

You can have the best PPE in the world, but if it is not put on correctly and removed correctly, it becomes a risk itself. There are two protective goals of PPE. The first goal is to make sure that you keep what you are working with where you are working with it. This means that PPE should be removed within the laboratory or health care facility to ensure that areas outside of the facility are not contaminated by "dirty" PPE.

The second goal of PPE is to protect all the body's portals of entry. Laboratory staff work with pathogens that, if they enter the body, can make them sick. Notice the word "if." To date, we have not found any infectious disease capable of penetrating a physical barrier placed between it and a portal of entry. If all portals of entry are protected, then there is simply no way for that pathogen to enter the body.

PPE may include gloves, negative-pressure respirators, positive-pressure respirators, surgical masks, booties, lab coats, Tyvek suits, lab-specific clothing, lab-specific shoes, eye protection, and sleeve covers. Each of these plays a key role in protecting the points of entry into our bodies. But it is important to remember that how PPE is used determines its overall levels of safety for the infectious disease pioneer.

It is important to note that PPE alone is not enough. PPE must be used with an SOP that has been evaluated, validated, and verified to ensure that the process used to remove contaminated PPE does not lead to the contamination of the infectious disease pioneer. A staggering 70 to 80% of laboratory-acquired and health care-associated infections cannot be traced to a known exposure event. My expert opinion is that most of these infections are acquired during the doffing of contaminated PPE.

Gloves

There was a time when babies were born into ungloved hands. Dentists would put their bare hands inside a patient's mouth during a variety of procedures. During the 1980s, with steep increases in hepatitis B and the emergence of HIV, a drastic change in the medical profession occurred that required professionals to protect themselves by wearing gloves during patient procedures. This change was readily accepted in research and clinical laboratories as well as the clinic, as it was clear that cuts and tears around the cuticles and on the fingers provided ample opportunities for bloodborne pathogens to enter the body.

Barrier protection is not the only benefit of wearing gloves. Gloves also permit easy removal when the user suspects that contamination has occurred, allowing the user to keep any contamination from spreading beyond the area of use. It is therefore critical that laboratory staff pay close attention to when and how they remove their gloves. I have witnessed thousands of laboratory staff, all over the world, improperly remove their gloves, contaminating their hands, and then, with those contaminated hands, fix their hair, adjust their glasses, rub their nose, and touch their face before washing their hands with soap and water. PPE is implemented to decrease risk. However, if staff are not trained in appropriate use and removal, then the gear intended to protect them becomes the gear that hurts them.

Negative-Pressure Respirators

Negative-pressure respirators work by creating a barrier between the mouth of the person using the respirator and the environment. Air is pulled through the respirator toward the user from the environment, hence the term "negative pressure." Although there are benefits to using negative-pressure respirators, such as preventing the user from touching his or her mouth and helping minimize the likelihood of droplet or other transmission via the mouth, there are some strong drawbacks to consider. There is no doubt that negative-pressure respirators provide ample protection against agents that are spread via the aerosol route when they are fitted and worn properly by staff who are medically cleared to use them. Smiling, laughing, sneezing, and having facial hair all affect the fit and the protection. Negative-pressure respirators are also uncomfortable, cause greater levels of fatigue in those who wear them, and can lead to chronic asthmatic conditions in those who wear them frequently over extended periods of time.

Positive-Pressure Respirators

In 2004, we opened the Emory University Science and Safety Training Program for BSL3 laboratories. Initially, we were planning to train with negative-pressure respirators but discovered that, even for training programs, respiratory fit testing and medical clearance were required for each participant, which was unrealistic for our program. To provide real-life scenario-based training on the use of respirators, we used positive-pressure respirators, which don't require medical clearance or fit testing when used for training.

In most cases, positive-pressure respirators provide the greatest amount of protection and comfort for laboratory staff. I would argue that, over a five-year period, positive-pressure respirators are more cost-effective from a maintenance standpoint than negative-pressure respirators. Negative-pressure respirators not only require the replacement of a respirator after a single use, they also require an annual medical screening and fit test for all infectious disease workers using them. Although initially more expensive, positive-pressure respirators do not need to be replaced over time, can be reused, and do not require annual medical clearance and fit testing of users, saving time and money.

Positive-pressure respirators provide the added benefit of built-in eye protection. When negative-pressure respirators are used, eye protection may fog up. Because this prevents the user from comfortably being able to see, staff will sometimes remove their eye protection or respirator, losing critical barrier protection. This does not happen with positive-pressure respirators, which are light and comfortable and do not allow easy removal, better ensuring compliance that protects all portals of entry in the face and head.

Surgical Masks

There is a misconception that influenza virus, HIV, and other agents transmitted through blood, fecal, and oral routes are spread the same way as agents such as the tuberculosis, anthrax, and plague bacteria, which are spread by aerosol routes. However, agents transmitted by blood, fecal, and oral routes are spread via droplet transmission, so the use of surgical masks or face shields provides protection. These are an inexpensive and effective barrier between the environment (clinical and laboratory), nose, and mouth and the hand. In other words, a surgical mask can ensure that what is on the hand cannot (through touch) enter the nose or mouth unless the mask is removed.

Sometimes staff working in a BLS2 laboratory will work in a BSC while wearing a negative-pressure respirator. If they are working with a pathogen spread by blood, fecal, or oral routes, in almost every case a surgical mask is a more than sufficient method of protection. The idea that laboratory staff in this scenario receive additional protection from a respirator is based in fear and lacks scientific evidence to support it. Upgrading PPE for the sake of sustaining an "industry standard" that was established without scientific evidence actually causes a reverse safety issue, putting the user at risk instead of preventing risk. For example, the use of PAPR (powered, air-purifying respirator) hoods for work done in a general laboratory with biosafety cabinets is excessive and expensive.

Shoe Covers and Booties

Although care is taken to make sure that we are keeping what we are working with in its appropriate space, things may still end up on the floor of the laboratory or health care facility. As we walk around, floor contamination may be tracked throughout the environment by the shoes we wear. Considering that many staff wear the same shoes in and out of where they work, this could quickly become a safety issue.

Animals in our homes may play with our shoes, and children may crawl and roll on the floor or ground. If this ground has become contaminated by something that was stepped on at work, an exposure event could happen. Wearing protective shoe covers and booties in the lab allows us to separate areas that could be contaminated from areas we are attempting to keep clean. Shoe covers and booties are a very good form of protection from any dangers that may be on the floor of the lab, but to provide 100% assurance, I strongly recommend using laboratory-specific shoes, i.e., shoes worn only inside the laboratory (see below).

Laboratory and Health Care Provider Coats

Laboratory and health care provider coats, when used properly, allow the wearer to maintain a dirty-to-clean workflow in a more efficient manner by protecting exposed skin and personal clothing worn to and from the work environment. I don't understand why so many labs place their laboratory coats near the lab exits. Usually I am told that is because the coat should be the first thing laboratory staff put on when entering. However, I disagree with this. The laboratory coat accompanies the wearer all throughout the laboratory, the arms move in and out of BSCs, and the coat may be exposed to agents worked with on the benchtop, so why place what has been potentially exposed to a pathogen near the entrance of the laboratory? Another possible contamination risk occurs when laboratory coats are stacked on top of one another. This behavior could potentially transfer contamination to the inside of a laboratory coat, which could contaminate the wearer's personal clothing. Instead, lab coats should be stored near the area of the lab where contamination could occur and should be donned after gloves to protect the bare hands of the infectious disease pioneers. Additionally, I don't understand why health care providers wear the same coats they treat sick patients in, to go to the hospital cafeteria!

It is true that most coats are not contaminated, and if someone knew that a coat was contaminated, it would be properly addressed, discarded, or decontaminated. However, because not everything can be seen, we don't have the evidence to determine whether a coat is clean or dirty. Common sense states that if you are working with something infectious, it would be wise to remove PPE that may be contaminated before going to cleaner parts of the laboratory or workplace setting.

Eye Protection

Eyes are one of the most direct portals of entry for agents spread via blood and droplet transmission. The BMBL states that eye protection in the BSL2 laboratory is to be used for anticipated splashes or sprays of infectious or other hazardous materials when an individual is working outside the biosafety cabinet. This recommendation implies that laboratory staff should wear eye protection only when they plan to cause a splash or spray, something I expect that staff never intentionally do.

Safety officials do their best to identify when and with what procedures eye protection should be used and to inform the staff on these guidelines. Despite this work, I have visited many laboratories in which the use of eye protection is sporadic at best. To ensure that the eyes are always protected, staff should wear eye protection any time they are working inside the laboratory, not just for specific procedures. This rule protects staff against splashes and sprays and also against their own hands, which frequently touch their face and eyes.

One important, but sometimes overlooked, aspect of eye protection is that it should be comfortable and personalized. Staff should not be forced to share communal eye protection, which is unsanitary, and staff should ensure that their eye protection has the best possible fit. Eye protection should not fog up, cause headaches, or get in the way of doing the work. Achieving quality eye protection is that simple.

This is not an exhaustive list of PPE but a summary of the common types used in most laboratory and health care settings. A thorough risk assessment, considering the overall layout of the laboratory or health care facility, should be used to determine the specific type of PPE needed. Decision makers should consider important factors such as staff comfort, affordability, availability, and overall ease of use (time and difficulty). The staff must be periodically trained in appropriate donning and doffing procedures and monitored for compliance.

STANDARD OPERATING PROCEDURES

The biological risk mitigation process consists of four phases. Hazards are *identified*, then each hazard is *assessed* and strategically *managed* with the development of SOPs. The final phase is *communication*, the explanation to staff as to why they are being asked to follow a specific plan while working in the laboratory or health care facility. You can expect large differences in overall behavioral compliance among staff in response to a behavioral request alone compared with a behavioral request that is paired with an explanation.

An SOP is two or more people doing the same thing, the same way, to achieve the same outcome. SOPs are a critical control for one of the biggest human risk factors. Staff will judge risk based on several factors, including their overall education and their experience with the risk. Attitudes about safety and overall safety behaviors are directly influenced by these judgments. We know that not all staff have similar education levels and experience, however. Within each laboratory or health care facility there are likely many different perspectives, and if we leave expectations for their own behavior up to the staff, there will be variations. With variations in overall behavior come variations in overall safety outcomes.

It is important to remember that an SOP is really a standard operating behavior (SOB). SOPs do not ask people to think or feel the same way; they require them to behave consistently around an identified hazard to ensure a consistent, safe outcome. I have witnessed the lack of SOP compliance in well-resourced laboratories around the world. Most people think they follow SOPs, but what people think they're doing and what they're actually doing are usually very different. When I enter a laboratory, I sit back and observe behavior. Generally, what I observe are many varied operating procedures, where two or more people do the same things but in different ways that produce different outcomes. Glove removal is a perfect example of this. Even in laboratories with resources and thick manuals, levels of SOP compliance are often low at best.

There was a time when I believed we needed a plan for everything, and the thicker the SOP manual was, the better and more prepared an organization was going to be. A small group of laboratory staff in the Honduras taught me that this way of thinking was wrong. While working with Anna Sanchez and her team at the Universidad Nacional Autónoma de Honduras in May of 2010, I noticed something unique. While observing behavior in one of the laboratories, I noticed…consistency. I saw true consistency among three different staff, all doing the same thing, the same way, and producing the same outcome—a very exciting moment! I asked to see their SOPs and expected to see a thick binder. They handed me a sticky note with a few items jotted down on it. This didn't make sense to me. How do you achieve consistency with no plan? The leader simply said, "We mentor and train them. Only when staff demonstrate how we want them to do their work do we let them do it." This simple observation changed my whole perspective on SOPs.

Behavioral consistency and SOP compliance are not achieved by writing plan after plan, placing them in a notebook, and having laboratory staff sign a page stating they have read and understood each plan. Handing staff a plan full of SOPs creates the premise that all SOPs are of equal value. This could not be farther from the truth. There are SOPs for how to write SOPs, and that kind of SOP does not compare to an SOP on how one should doff PPE when leaving the laboratory or health care facility. A spill cleanup SOP is far more important, in most cases, than how one enters the laboratory or health care facility. Placing SOPs one after another in a book dilutes the critical SOPs—the ones that directly impact safety—with the noncritical SOPs. If you are an experienced staff member, you can likely review an SOP manual and distinguish between what matters and what does not. However, if you are new to the laboratory or health care facility, you may give the same amount of weight to an SOP about separating laboratory or health care facility waste properly, which has a minimal impact on safety, as you do to doffing PPE, which is safety critical.

Unfortunately, most SOPs are not written for the proper audience, which is the workforce. The fact is that, although SOPs are developed for the workforce, they are presented to and approved by leaders and regulators and are meant to demonstrate adequate and strategic risk management processes for protecting laboratory staff. Leaders and regulators have the luxury of reviewing SOPs from the sidelines; their reading and reviewing of SOPs doesn't affect their overall health and safety risk. However, those being asked to follow SOPs must go beyond understanding an SOP for it to be effective: they must execute it.

The failure of SOP controls usually occurs because of how the SOPs are carried out by the workforce. A plan by itself will never produce consistent behavioral practices among different people, just as a recipe card alone does not make a cook into a chef. Laboratory and clinical staff must go beyond reading the SOPs: they must be mentored and trained to consistently demonstrate the desired behaviors. SOPs should be integrated into workforce behavior by means of effective mentoring and training techniques, so staff can execute the behaviors.

This is where I have introduced the "SOP-free" training model. The problem we face today in both laboratory and health care settings is not the SOP itself but how the SOP is carried out by the workforce. Usually staff receive copies of the SOPs and are tested on their reading and comprehension skills. However, if staff were trained without SOPs (meaning that they don't have a hard copy) and we focus on their behavioral compliance with the SOP, training programs would not only be more effective, they would also verify that staff can do what is expected of them specific to the SOP.

Remember, an SOP is written for a specific behavior, and if that behavior varies even a little, it drastically increases the risk to health and safety. Do all your SOPs measure up to this standard? If not, perhaps you have included standard operating *policies* in your safety manual. Another thing to distinguish is the difference between safety SOPs and scientific SOPs, which have different purposes. Scientific SOPs are reviewed for safety concerns but are primarily developed to ensure that safe science can be replicated from one laboratory or health care facility to another. Safety SOPs focus on replication of safety among different staff within a specific laboratory or health care facility. My point is that if variations of behavior do not substantially increase overall risk to health and safety, this is likely a policy. Save the term "SOP" for behaviors that must be done consistently to preserve an acceptable level of safety.

The acronym SOP should be powerful. SOP should signify risk if variance of behavior occurs. Calling everything an SOP weakens an opportunity to send a clear message to staff. An SOP for how to write an SOP should not in any way resemble an SOP for doffing PPE. I urge you to create a *policy* for how to write an SOP, and write, review, approve, and practice an *SOP* for how to doff PPE. SOPs are sacred to health and safety of staff and should be unique and respected.

RATING YOUR SAFETY PROGRAM

The four primary controls provide safety officials with the opportunity to serve an organization right where the organization happens to be. For example, in the United States, most organizations tend to rely heavily on engineering and PPE controls, writing SOPs for regulators and dabbling in administrative controls to check specific boxes, such as training, vaccinations, background checks, and occupational health clinics. On the other hand, in countries that do not have the resources to focus on engineering and PPE, they focus on SOPs and leadership controls, ensuring that what they do with what they have is done properly. With this said, I believe there should be an easy yet effective manner in which to evaluate overall safety programs.

For several years, I have evaluated safety and clinical containment programs by asking leaders, safety officials, infectious disease pioneers, and outside consultants to rate their overall safety program using the four primary controls and providing justifications for each rating. Each primary control is rated using the following scale: A = excellent, B = good, C = fair, D = poor, and F = unacceptable. Based on all of the answers I have received over a ten-year period, the common trend in the United States is for these individuals to rate their programs similar to the following.

1. **Engineering = A (Excellent)** Most research laboratories and health care facilities in the United States have world-class facilities, almost to a fault. Typically, the more money you spend on engineering, the more money it takes to sustain the facility. This applies to both laboratory and health care facilities. (However, public health and clinical laboratories within hospitals need serious engineering upgrades to keep up with emerging infectious disease threats. I would give them a C+ at this point.)

2. **PPE = A (Excellent)** Facilities in the United States usually have access to plenty of PPE. Even the Texas hospital which had the two nurses become sick with Ebola had the best PPE available. It is a

humbling experience to use more gloves in one training course given in the United States than some staff have allocated for a whole year in under-resourced countries. The only reason I have seen PPE in the United States receive a lower grade is because the PPE is uncomfortable, breaks, or is impractical. Safety leaders should listen to complaints from staff using PPE. If they don't like the PPE, they will be less likely to use it.

3. **SOPs = C (Fair)** Right now we have SOPs for how to write SOPs. Many organizations today provide the illusion that they are updating their SOPs when in reality they are simply signing off on the SOP annually. SOPs are rarely evaluated, validated, and verified. Many evaluators complain about how SOPs are written, the number of them that exist, and the fact that there is usually enormous difficulty in changing the SOP.

4. **Leadership = D (Poor)** We have a serious lack of leadership in both laboratory and health care settings. Expectations are set, resources are provided, and then what? We wait until something goes wrong, blame the staff, fire or reprimand them, and move on. Leaders must get engaged. They must promote safety among their staff by being involved and providing feedback to staff about their safety behaviors. They must ensure that staff have the resources and training needed. They must nurture an environment where mistakes can be reported. Leaders must address those who are not complying with SOPs and protect those who are trying but falling short. Leaders cannot lead from the top when dealing with infectious diseases; they need to be in the trenches as champions of the infectious disease pioneers. Right now, the chef has left the kitchen, and cooks are running the show. Get out, get about, and lead!

When looking at how under-resourced organizations work with dangerous pathogens and patients safely, here are the current ratings I have averaged from those serving in such countries.

1. **Engineering = C (Fair)** Let's face it, not every country has the money to purchase millions of dollars' worth of equipment and maintain it. I have traveled to laboratories in many countries that have been given thousands of dollars' worth of equipment, and it is collecting dust. Good, outstanding, cutting-edge equipment, just sitting around. Why? Because they don't have the training, resources, or ability to maintain it.

2. **PPE = D (Poor)** One of the leading killers specific to infectious disease is tuberculosis. Imagine working with this agent and patients sick with tuberculosis and having no gloves or respirators. The inability to regularly supply staff with PPE is a real problem in under-resourced countries. Someone may be asked to work a whole day using only one pair of gloves. As a reference, most staff in the United States report

using approximately 10 pairs of gloves a day. Health care providers report using even more.

3. **SOPs = D (Poor)** Countries that are under-resourced have many workers who have been educated in well-resourced countries. This is an outstanding opportunity for them; however, when they return home, they find that the practices of the well-resourced country are unrealistic in their home country. Attempting to convince others to create and follow SOPs is equivalent to asking the *Titanic* to turn *now!* SOPs may be present but are very difficult to integrate and implement into the workforce.

4. **Leadership = D (Poor)** Although leadership controls can be implemented in innovative ways to save costs, plans and solutions to do this have not been developed. Absentee leadership is a pandemic. I fear most leaders today are assigned leadership roles but are psychologically absent from this role. This allows others to disengage from safety initiatives altogether. Dave Franz, a great leader, during his assignment at the U.S. Army Medical Research Institute of Infectious Diseases pulled safety out of the basement and onto the executive level. When leaders say safety matters, it begins to matter.

Notice that the averages of these ratings do not quite equal a B (for either)—more like a B– or C+ in the United States and a D in under-resourced countries. This should cause some concern in those reading this. Although these averages are not statistically evaluated and validated, I have quantified (to the best of my ability) how programs are rated and find truth in the ratings provided above. Additionally, the United States doesn't necessarily attract the most dangerous agents in the world. All countries should have the capability to prepare for and fight emerging infectious diseases, because pathogens do not acknowledge international borders. We are one world.

In summary, we should never rely on one primary control more than another; all four are equally important and should be treated as such. I recommend evaluating safety programs with perception surveys on a quarterly basis. Though engineering and PPE controls are typically financially driven, SOPs and leadership controls can be implemented even in places where financial support is weak. If your organization doesn't have the resources to increase your engineering and PPE controls, focus on the SOP and leadership controls to balance the weaknesses in your safety program. Using a rating technique to examine existing safety efforts provides insights, based on perceptions of the four different controls, that could result in a healthy and substantial list for safety officials to use when looking to improve overall safety programs.

How would your safety program rate today? Remember, perceptions influence safety attitudes and behaviors, so ask your people what they think. In my experience, not only is this an exercise that identifies proactive strategies for minimizing risk, it is one that produces trust, surprises, and unexpected changes in the overall program approach.

BIOSAFETY *in the First Person*

The Dancer of Biosafety by Jim Welch

Jim Welch knew Beth Griffin's family and was her teacher in middle school. Here, he explains what led a history teacher to enter the profession of biosafety and lead a charge internationally to increase awareness and ensure proactive biosafety approaches within many organizations. I have always appreciated Jim for many things, including his humility, his ability to connect with people, his friendship, and most importantly, his dedication and commitment to safer science. He is so dedicated that, in spite of being a tall guy, he has traveled the world in economy class, assuming practically a "cannonball" position for up to 24 hours. I bring this up because Jim uses one of my favorite analogies. On airplanes, there is a difference in safety approach between first class and economy class. When the seat belt chime comes on, those in economy class are typically monitored and corrected for standing or going to the restroom. In first class, however, it is a different story. Jim reminds us that safety must be applied to everyone equally and that in laboratory and health care environments, there is no first or economy class seating.

My first career was as a middle school history teacher in a public school in northeastern Tennessee. As with most new teachers, I had an anticipation of making a significantly positive impact in the lives of my students. I just knew that my love of the intricacies and nuances of history would be so contagious that the phrase "I hate history" would never be muttered by any student who ever graced my classroom. Little did I know that it was my life that would be dramatically changed by one student.

Teaching middle school is much more about managing puberty than it is about academic growth. Successful middle school teachers learn quickly to slide academic learning into whatever opportunities may present themselves. That which totally dominates the life of the average middle school child is how they look and what other people think about them. I used to jokingly say that the meanest group of people on the planet is a group of middle schoolers who decide to turn on someone. They are merciless, and their choices of who is out and who is in change as often as the wind. In my 30 years of teaching middle school, I can count on one hand the students I had who were obviously above that. One of those was Beth Griffin.

Beth's family was very active in the community. Her father was a well-known physician and her mother was both a community leader and a very successful health care management professional and consultant. Their older daughter, who is now a physician, was a dedicated student and ran cross-country. Beth was known for her dancing prowess. She excelled far beyond what was normal for our area, and it was so much a part of her that she actually danced when she walked.

What floored me about Beth the middle schooler was that nobody, and I do mean nobody, hated her. More than that, I never saw her participate in any of the ongoing social rituals that made people move from their seats in the cafeteria, laugh at others about how they looked, or fail to be friendly with anybody. While every bit as smart as anyone would expect her to be, she enjoyed far more in life than getting awards for excelling in class.

I kept up with Beth and her sister through her parents, as we were active in the same church. Around the same time Beth was finishing her studies at Agnes Scott College, her mother answered a call into full-time ministry and began studies at Candler School of Theology at Emory University in Atlanta, Georgia, the same town where Beth took her first "grown-up" job with Yerkes National Primate Research Center.

I remember attending our church Sunday school class in October 1997, where a prayer request was raised for Beth. It seems as though she had contracted something from a monkey via a splash in the eye and was having some medical issues as a result. I also remember that part of the story was her having difficulty getting appointments with physicians regarding her problems because somebody somewhere had deemed that no follow-up was necessary. We collectively nodded and sighed as people often do with such news and life went on.

That changed quickly and dramatically. We learned that Beth's situation was worsening. By the time the third week came around, it appeared as though everything that could possibly go wrong had done exactly that. We started learning about a virus that macaques live with and can shed that can actually kill human beings. It suddenly hit our community that this beautiful person who danced and danced was losing her mobility below her neck.

For some inexplicable reason, I became prayerfully convinced that everything would be okay. I had a spiritual certainty that some miracle would happen, and Beth would be healed and fully recovered.

I was just as wrong as the person who declared that no follow-up was necessary. I was just as wrong as the diagnosis of cat scratch fever. She died a horrible death. Her family and friends were devastated. The church in her home community overflowed at her memorial service. Her death became national news and the focus of a story on an ABC news program. Some people's deaths strike us especially hard, and this was one of those times.

In the aftermath of their loss, Beth's family set up a foundation to honor her. The purpose was to improve occupational health and safety awareness for people who work with macaques. For some unknown reason, the Griffins asked me to be its part-time executive director. They wanted me, a middle school social studies teacher, to reach out to groups to let them know our mission was to work collaboratively with them to help make their work safer, not to eliminate their work.

The first group we reached was the Association of Primate Veterinarians. Amazingly, it is these very people who frequently are on the cutting edge of medical research. When the world responded to the severe acute respiratory syndrome outbreak in 2003 with a rapid global expansion of high-risk laboratories, the Association of Primate Veterinarians encouraged us to carry Beth's story into the larger world of biosafety, and we did exactly that with the help and encouragement of the American Biological Safety Association.

I was blessed to spend 17 years professionally devoting my efforts to promoting safe science and working toward a goal of no more tragedies like Beth Griffin's. In that work, I have had the opportunity to learn from and work with highly dedicated, highly motivated, and highly capable people around the world. In my heart, I know that the young woman who danced everywhere she went is still dancing. You can watch her dance every time you witness safe behaviors in laboratories, and you are always invited to dance with her.

chapter 6

Understanding Human Behavior

While conducting biosafety training in Pakistan, I decided to do an experiment. I split the large group of participants into two groups. I asked the groups to rate their moods using a simple mood-rating scale: If they were in a good mood, they were to check the smiley face. If they were in an okay mood, then they were to check the normal face. If they were in a bad mood, they were to check the frowning face. Here are the results I received.

Group A			Group B		
Good	**Okay**	**Bad**	**Good**	**Okay**	**Bad**
9	5	0	6	8	1

I asked group A to leave the auditorium and step into the hallway. After group A left, I asked group B to show me their cell phones (all but one participant had a cell phone) and then asked them to text a loved one, telling that person how much they were loved and appreciated. This took about 10 minutes to complete. Then I asked group B to again rate their moods (hiding the previous results). Finally, I invited group A to return to the room and again rate their moods (hiding their previous ratings as well). Here are the new results.

Group A			Group B		
Good	**Okay**	**Bad**	**Good**	**Okay**	**Bad**
9	5	0	14	1	0

What happened here was not too surprising to me. Within a 10-minute period, I was able to change the mood of one group by asking them to participate in a specific behavior. I like to use this technique to prepare a group for a training session because the better people's mood, the more receptive they are.

Human behavior is a passion of mine. In 20 years of applied behavioral training practices, I have witnessed and discovered behavioral trends among individuals that do not vary from country to country or culture to culture. Rather, the responses to a similar stimulus in similar environments are instinctive, predictable, and quite reliable regardless of where you live, the language you speak, or your religious beliefs. My education, training, and experience have led me to believe that the blending of behavioral and cognitive psychology produces the greatest understanding of what people will do when they encounter specific situations.

Was my experiment about changing mood or about changing behavior? Either way, the group who was asked to express love and appreciation to others reported feeling better afterward than the group who was asked to stand and wait in the hallway. In this chapter, I will explore how we influence both behavior and cognition when we set expectations and why this is critical in achieving safer behaviors.

WHAT CAUSES BEHAVIOR?

There are people far more qualified than I to answer this question. No doubt my explanation will simplify what most of those individuals might have to say, and they may even disagree with me. Having served alongside people who are experiencing a crisis, whether as a result of an infectious disease, airline crash, terrorism, refugee status, or life in general, it's clear that humans have basic needs. I believe that meeting those needs is what drives behavior. Without need, there is no behavior. It is that simple.

Humans are at the top of the food chain for many reasons. One of those reasons is that we are very efficient with our personal energy. If there is no need to behave a certain way, we don't. Even when there is a logical need for a certain behavior, such as to reduce our use of environmental energy resources, our personal needs take priority over the environmental ones. Each behavior has a reason. If someone's behavior does not make sense to you, maybe you need to look more closely for the reason.

Maslow discussed four basic human needs and placed them in a hierarchy. Let's flatten that hierarchy and imagine that the needs are equal. Let's ask ourselves what happens when each one of these needs is challenged.

- The need for air, food, water, and shelter.
- The need for safety and security of body and mind.
- The need for belongingness.
- The need for self-value and esteem.

If there is a challenge to any one of these needs, the response is behavior. I have asked people around the world why they brush their teeth. Here is what they tell me:

- I brush my teeth to prevent decay. If I lose my teeth I cannot eat. (physiological need)
- I brush my teeth because I am afraid of the dentist. (safety and security)

- I brush my teeth because I don't want others to be offended by my breath. (belongingness)
- I brush my teeth because I like my smile. I like the way I look with teeth. (self-value/esteem)

We can do this exercise with almost any behavior and determine the reasons behind specific behaviors. There is not just one reason for brushing your teeth but hundreds that are based on meeting our basic needs.

To understand human behavior, we must look for the reasons behind it. I want you to consider and accept the concept that all behavior occurs because of a need to behave that way. This is the first step in understanding behavior rather than judging behavior. So, what drives the need to behave?

WHICH IS STRONGER: YOU OR THE ENVIRONMENT?

Basic behavioral psychology states that a person's *perception* of risk will influence his or her *attitude* (desire) to *behave* (respond). In short, perceptions drive attitudes that drive behavior (P→A→B). But we need to consider the factor of *control*.

There is a battle that occurs in humans, a fight between what someone thinks and feels versus what they do. If an individual has control over the situation—the ability to choose which behavior they will participate in—it is easy to do what he or she thinks and feels. When these things align, the individual is not conflicted and is able to rationalize risk and innovate improvements within the individual environment. Humans naturally strive to achieve this alignment, simply because it takes tremendous energy to live in a conflicted state. A conflicted state occurs when we are forced to behave in a manner that does not match what we think or feel. This happens when our environment, and subsequently our individual behavior, is out of our control.

If people do not have control of their environment, or the people in their environment, they may be forced to accept things that they do not initially agree with. I say "initially" because if they truly do not have the ability to change the environment and cannot leave the environment, in time it becomes much more efficient to accept the environment than to resist it. Humans may not be efficient with resources, but when it comes to energy exerted for purposes of behavior, efficiency prevails. The environment wins, and it begins to drive what a person thinks and feels to justify the behavior.

Now, how does this affect laboratory and health care environments? If staff have control over their environments, they will choose individual behaviors based on their personal perceptions of risk. At the risk of dismissing the importance of individuality, when it comes to dealing with humans, behavior, and risk, how a group of individuals choose to interact with the risk is of vital importance.

This is where the concept of *safety culture* comes into play. The backbone of any type of culture is compliance with common rules that protect against common risks and create common rituals among a specified group of individuals. If we allow individuals to choose their behaviors, we know perceptions of risk will vary drastically. These variations will produce differences in desires to behave, and as a result, there will be differences in behavior. Different behaviors produce different outcomes, and this is the issue.

Good scientific research can be replicated, no matter who is doing it or where it is being done, as long as it is being done in the same way. Safety must subscribe to the same standards. If we allow staff to use their own perceptions of risk to choose their overall behavior toward that risk, we will not be able to produce consistent safety outcomes among different individuals in different locations.

Instead of focusing on the most obvious risk, the biological agent, our focus must be on the behavior of individuals working with the risk. If their behavior is not consistent with the cultural practices, it must be acknowledged, addressed, and modified. Removing choice minimizes variations in behavior that result from differences in levels of experience and education among the workforce. To achieve consistent behavioral practices, having the choice of how to behave must be removed. Although this sounds harsh, it is already happening.

Standard operating procedures (SOPs) are defined as two or more people doing the same thing, the same way, to achieve the same results. They standardize a process, allowing consistent outcomes to be achieved by different individuals. In short, SOPs change the dynamic from perceptions (P)→attitude (A)→behavior (B) to B→A→P. When choice is removed, it is behavior that begins driving attitudes and perceptions. There are historic examples where society was given a choice and then the choice was removed, and as a result, change occurred. For example, consider car seat belts in the United States. Before having seat belts in every car became a legal mandate, many people did not believe they made a safety difference. However, after choice was removed, we adjusted and accepted them. Now most of us would not consider driving without our seat belt fastened.

Expectations of how we should behave in health care and laboratory settings serve as the environment to which we must cede control. The need to ensure consistent behavioral practices among different individuals must be stronger than the individual's choice of how to behave within these environments.

WHAT IS NEEDED FOR HUMAN BEHAVIOR

It should be clear that people behave for a reason: to satisfy their needs. You may not like or understand the reason, but knowing they have a reason gives you an opportunity to see the behavior in a new light. When the goal is to change a behavior, if you do not address the need that drives the behavior, the need will continue to exist and the undesirable behavior will persist.

It's also important to understand that human behavior is influenced by internal and external factors, and whichever is stronger will dictate either individuality or consistency. If individuals are able to choose their behavior, their individual perception of risk plus their overall experience in dealing with risk will drive the behavior. Unfortunately, this leads to high levels of behavioral inconsistency. However, if the environment controls behavioral choices, consistency among different individuals can be achieved.

A major factor that influences compliance is access to what is needed to behave as desired. One of the worst things we can do to another human is communicate that a risk is present, convince them they should avoid that risk, and then give them no way to do that. This leads to apathy—cultural acceptance that a high risk factor is the norm—which can be globally devastating to human health.

Understanding this basic concept about human behavior can help you in your goal to achieve safer behaviors. Have you ever seen advertisements offering seminars in personal growth focused on teaching you how to make money in short periods of time? During my graduate school days, extra credit was offered for attending a seminar that promoted something like this and then theoretically dissecting what was happening and what was missing. The seminar I chose to attend was about becoming a millionaire in one year or less (or some similar title).

The speaker was a self-proclaimed millionaire. He was enthusiastic and energized, and I became convinced that he believed every word he said as he said it. He began by discussing the risks of not being a millionaire. His research was clear: if you did not have money, typically you were not well educated, died at a younger age, and lived a more reclusive life. The benefits of being a millionaire were living a longer life, a higher level of education, and more opportunity to see the world, as well as lower levels of depression and anxiety and even a better sex life. That caught my attention. More money meant better sex? Hmmmm….

There was a break halfway through the seminar. When we returned, he turned his attention to convincing us that we could do it—become millionaires in one year or less! He talked about real estate, investing, entrepreneurship, and banking, asking us to visualize the cars we wanted, the dates we would take our loved ones on, and the houses we would own. He asked us to make a wish list of things we wanted and decide what we were willing to do to achieve those things. He assessed our interest in each strategy, and my greatest interest was real estate. Looking back, by the time I left the all-day seminar, I had some ideas about how I was going to become a millionaire in one year.

When I realized what had happened, I quickly returned to reality. The strategy was brilliant and is, in fact, one that all of us have likely used. When talking to our children, partners, spouses, patients, or coworkers about behavior, we create a need to behave by discussing a specific risk. In the wealth seminar, the risk was living poor. Specific to safety, we will also focus on the benefits of behaving as we are asking other people to. The seminar facilitator focused on benefits of being rich, selling the idea of wealth as a solution to the risk of being poor. If we are successful in communicating the risk and selling the benefit, then we can get people to believe they can do the behavior, right? Wrong!

There was something missing in this seminar. The seminar had certainly identified risks and benefits. It even convinced me that I could do it! But it did not provide me with the resources and skills I needed to succeed. And so, after nearly 5 years, I have not become a millionaire, unfortunately. Even though I clearly understood the risks and benefits, and even believed in myself, without the needed resources or skills it did not happen.

Just because humans have a need to behave does not mean they have the capability. Laboratories and hospitals may establish expectations, but if they do not provide the things needed for sustained behavior to occur, it simply will not occur. This is where leadership becomes essential. A leader asks the question, "What did I fail to do when someone has not done what I have expected him or her to do?" Successful leaders ensure that those they work with have the items needed to behave as desired.

There are five things that are essential for sustained behavior to occur. Consider these each time you create, train with, or discuss your SOPs.

ACKNOWLEDGING SAFER BEHAVIORS: PUNISHMENT VERSUS REINFORCEMENT

When behavioral requests are being made by organizational leaders, it is important that the leaders recognize whether those expectations are being met. Remember, humans are efficient with their energy. Why would they behave a certain way if they don't have to? Failure to acknowledge and hold people accountable for those behaviors minimizes the likelihood that behavioral expectations will be followed. If you are expecting humans to behave and you fail to acknowledge whether the behavior has occurred, it was never an expectation—it was a hope. Hope is an expensive commodity when we are dealing with the interaction between human behavior and the risk of working with infectious disease.

There are two types of accountability. Positive accountability focuses on acknowledging when someone is doing something as expected. Negative accountability focuses on acknowledging when someone is not doing what is expected. It is unfortunate that most safety training focuses on negative accountability. This creates a state of policing rather than coaching.

Police officers use punishment as a form of negative accountability. The jury is out on the effectiveness of punishment regarding behavior. When someone is being policed, they are being considered through the filter of what they are doing wrong. The goal of punishment is to stop a behavior, so the emphasis of those policing is on identifying all behavior that must be stopped. In truth, you cannot actually stop behavior, but you can stop it from happening around the person doing the policing. To many, this appears as though the behavior has been changed. But when the biosafety professional arrives, people eating in the laboratory hide, those wearing sandals run, cell phones are turned off, and those without a laboratory coat quickly put one on. This is very similar to speeding on the freeway. You see a police officer, and you slow down. The police officer pulls off the road, and you speed up. By policing, you are not actually stopping the behavior, just hiding it. This is an ineffective approach, one that increases risk. The less the safety official knows about true behavior, the greater the risk is to those serving the organization.

There is a value to policing in any society. It provides tremendous societal benefit in terms of satisfying human needs of safety and security. Holding people accountable to the behavioral expectations, i.e., enforcing the laws, is critical in maintaining this sense of safety and security at a community level. Is there a way to police more effectively?

What if more time was spent emphasizing what people are doing well and correctly, rather than what they are doing wrong? I know it seems like a naive idea, but let's make some assumptions. What if risk equals the gap between what is seen and what is really being done? How would police officers be able to minimize risk if the people they intend to serve are not helping them bridge that gap? For instance, if police officers establish trust and form a relationship with those they serve, will the community not choose to assist in bridging the gap by disclosing that certain individuals are saying one thing while doing another? Anyone who knows anything about police work knows this. The more the community helps, the better the result. Why on earth would anyone help someone who always sees the worst in them?

If you want to build trust and form relationships, you must focus on what someone is doing right. This is positive accountability. In collaboration with the American Society for Microbiology, my company offers a program called 52 Weeks of Biosafety to support safer work environments. One aspect is that, each month, participants are asked to identify a person in their organization who is participating in a safer behavior. "Safer" refers to a behavior which is in compliance with the organization's guidelines, which was developed as a result of a risk assessment and strategic risk management approach. They are asked to acknowledge the safer behavior, show appreciation, take a photo of the behavior, and submit it as their completed assignment. This is part of an effort to train individuals not only to see what is right but to acknowledge it and appreciate it. This process fosters trust and builds the relationships needed to bridge the gap between what people say they do and what they really are doing when nobody is around.

If policing hides true behavior and coaching focuses on true behavior, which would you choose? I tell everyone that if a specific behavior is an immediate risk to life and safety, it must be stopped immediately. However, that is not all that is needed. The risky behavior must be replaced with a safer one. If the new, safer behavior is easier to do than that old risky behavior, then change can occur quickly, because humans will follow the path of least resistance.

Pointing out a risk and providing resources to people will never be enough to motivate safer behavior. We must go further by setting safer behavior expectations and holding people accountable for them. When holding people accountable, start with recognizing effort and what they are doing well. Make a request regarding items needing improvement with a question, "Is there anything you need which you do not have to achieve this behavioral expectation?"

CONCLUSION

I have so much to share on this subject. I could discuss multiple theories I utilize to determine whether someone is ready to change behaviors. We could explore the stages of learning individuals follow. Instead, I challenge you to prioritize the behavior of staff in the laboratory and clinical setting as the greatest risk of all. What does that mean?

It means that compared to an emerging infectious disease, your behavior on the job serves as your greatest risk not only to health and safety but to whether you have a job or not. During graduate school, I was taught the difference between health education and health promotion. Health promotion is always more successful than health education because of one word: choice. Health education provides information to people and allows them to choose what is best for them. Health promotion removes choice by creating laws that protect individuals and the community, and because they provide accountability when broken, they lead to the greatest amount of social change. Some examples of laws that are health promotion include those governing smoking in public places and use of car seat belts.

To simply provide health education increases awareness but does not necessarily increase safer behaviors. We must move from health education programs to health promotion programs. We must understand that need drives a variety of behaviors. We must realize the environment is the greatest influencer of group behavior. Leaders must ensure that people have what they need to behave safely, and leaders must hold staff accountable (positively and negatively) to behavioral expectations.

BIOSAFETY *in the First Person*

It Could Have Been HIV by Henry Mathews

Henry Mathews has been one of my biggest mentors and supporters. Henry has spent his life serving those working with infectious diseases, as his wise demeanor and whitened beard show. I frequently thank him for all he has taught me, and true to his humble personality, he always responds that I have taught him more than he has taught me. He spent nearly 25 years working in the laboratory as a scientist at the CDC before transitioning into the biosafety field. Henry talks about some of the important lessons he learned during his time as a researcher about how behavior links to biosafety. Thank you, Henry, for your lessons, for your service, and most importantly for your commitment to the profession of biosafety.

Long before I became involved in biosafety and long before the BMBL or the Bloodborne Pathogen Standard, I learned some valuable lessons that I have carried with me for the rest of my career, both in the laboratory and later as a biosafety practitioner. In the mid-1970s, I was responsible for a malaria serology laboratory. We tested specimens from locations all over the world where malaria was endemic. My lab was tasked to participate in a large study of children from West Africa. Sera were collected and split into multiple vials for testing in my lab at CDC as well as labs at NIH.

My lesson began in the field, due to how the samples were collected and prepared, which was out of my hands, and other lessons were to follow. Let me explain. Sera were placed into Nunc cryovials, which had a silicone gasket to seal the vial for freezing. Specimen numbers were typed on paper (when people still used typewriters) and then cut out and attached to the vial with cellophane tape. To prevent leakage, I suppose, someone tightened the caps so tightly that the silicone gasket was deformed, then the label was attached with the tape. The tape was wrapped securely around the vial from the very bottom to the very top. The vials were frozen and shipped off to the participant labs.

When we examined the samples on their arrival, it was obvious that many of the vials had leaked. The cellophane tape was coming loose, and there was a risk of losing the labels. Because I could see that handling the leaky vials was going to be messy, I recommended that my coworker and I wear gloves. At that time, wearing gloves was not a common practice in many labs working with human serum samples.

As we began getting ready for our first day of testing, we quickly discovered that handling the thawed, leaky vials with gloves was almost impossible. The glue from the tape had separated and clung to the gloves, making it hard to put the vial in a rack without losing the label as we freed the vial from the glue adhering to the glove. After trying several options to open the vials, keep the labels in place, and avoid the glue on the gloves, I made a foolish mistake. I chose to remove my gloves and remove vials from the freezer one or two at a time. With a scalpel, I cut the frozen tape away from the cap. I insisted that my coworker continue to wear gloves, and I chose to wash my hands frequently. In some cases, we needed to retape the labels, and we were careful to recap the vials without distorting the silicone gasket. We finished our testing and returned all the samples to the freezer. I felt very clever when I was contacted by an NIH laboratory to see if I could share a subset of the samples, because they had lost all the labels when they placed samples in a water bath to thaw. We were able to provide the requested samples, labels intact and caps properly installed.

Because the study was long term, and I would not be involved in data analysis, I soon put the experience behind me and continued to test specimens, largely glove free. Life was fine, and about two months later, my wife and I along with another couple went out to a steak place to celebrate my birthday. For several days before my birthday, I had been feeling under the weather, and the thought of a good steak dinner seemed like the thing to perk me up. It didn't. I continued to feel bad, and it wasn't getting any better. A couple of days later, I was washing my face when I happened to notice that my eyes were a classic shade of jaundice. I knew immediately that I likely had hepatitis. But what kind? I had recently traveled out of the country and could have been exposed to hepatitis A. Or could it be a laboratory exposure to hepatitis B? Based on incubation times and lab testing, hepatitis B infection was confirmed.

The incubation time coincided with the leaky vial study, so it was time to review what had gone wrong. It had to have been skin exposure, but I was very careful with the scalpel and I knew I had not cut myself. Did I wash my hands often enough? I'm kidding myself, right? Then it dawned on me: I had been refinishing furniture during the time when I was working with the leaky vials, and I'm sure I had cuts and splinters from sanding old oak chairs.

To complete the hepatitis part of the saga, I was out of work for several weeks and then returned to a load of paperwork requiring me to explain why I had not reported the "accident" when it happened. I was the target of a laboratory-acquired infection investigation. I have fully recovered, and the last time I was checked, I was hepatits B antigen negative.

Here are the lessons that I learned.

Know the nature of the equipment you are using. In this case, the vials selected are widely used because they are sturdy, have a top that seals (if applied properly), and can be labeled with permanent markers. Many of the problems we faced could have been prevented if the vials had been properly prepared. On a larger scale, much equipment in today's laboratory has the potential to cause harm if not operated properly. Autoclaves, for example, may not decontaminate waste if improperly loaded and operated. Also, personnel can be burned or injured in other ways if the autoclave is used improperly. Centrifuges can be very dangerous if users are not well trained. Centrifuge failures can cause physical damage and can release infectious aerosols when mistakes are made. There are countless materials in the lab that require training so they can be operated properly.

Packaging and shipping must be done properly. In our scenario, we noticed that the cold chain had failed because it was clear that vials had thawed and had been refrozen. For the kind of serology we were using, this was not an impediment to accurate results. Freeze-thawing can be devastating to many biological products that must be transported nationally and internationally. Packaging and labeling requirements are much stricter today for carriers to accept them.

Personal protective equipment (PPE) must be used and used properly. The Bloodborne Pathogens Standard has driven home the requirement to wear gloves when handling potentially infectious material. Staff must be trained to don and doff PPE properly to avoid contaminating themselves or their surroundings. I need hearing aids today because I failed to use hearing protection in the lab as well at home with lawn equipment. Handwashing deserves some basic training and is no substitute for gloves. Believe me.

Don't hesitate to contact specimen submitters when samples arrive in less than acceptable condition. Had I done this, maybe some of the issues could have been corrected. In a similar vein, don't hesitate to call out a coworker who is not following standard operating procedures (SOPs). SOPs are in place to reduce exposure of individuals and contamination of the environment. It takes only one.

An infectious agent needs a susceptible host to cause an infection. There are vaccines available for many of the agents encountered in biomedical labs. There was no hepatitis B vaccine for me in the 1970s, so I got my titer the old-fashioned way: I endured the infection. Immunization is much better. Although vaccines may not be 100% effective, they can at least moderate the seriousness of an infection.

Universal precautions must be understood and practiced always. It is easy to be lulled into the careless idea that the sample contains only the infectious agent you are expecting. I was thinking malaria, not hepatitis B. At the time I was working with these specimens, there was an increasingly prevalent but yet unidentified agent, HIV. If HIV had been in those vials, I would have been a candidate to be the index case!

chapter 7

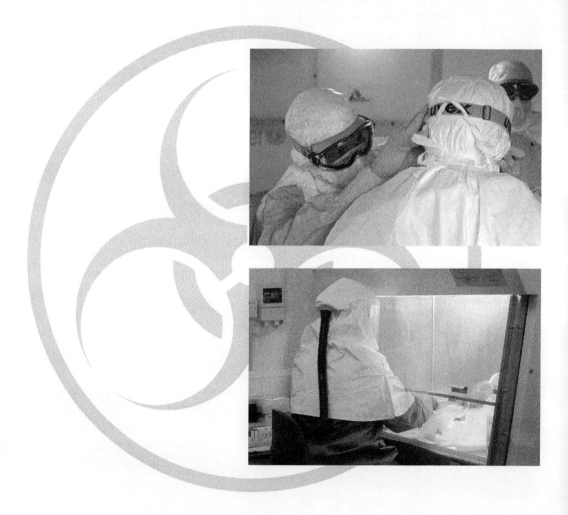

The Containment Philosophy

The containment philosophy is the idea of keeping something harmful under control or within limits. I propose that "something harmful" includes the infectious disease agent (in patients or vials), the people working with the agents, and the culture of the organization where the agents are being worked with. Instead of covering every possible issue surrounding containment, I will highlight some of the critical actions that keep what is harmful within the agents, people, and organizations under control.

The process of containment must start with the control of the agents. You cannot know what is harmful unless you know what you are up against. If you are a research scientist, you typically know what agents you are working with. However, if you are a clinical laboratorian or a nurse working with a sick patient, you are processing samples and placing hands on patients hours before knowing if what you are working with can cause substantial harm to you or others. This fact alone calls for practicing the containment philosophy to control the agent using the four primary controls of biosafety—engineering, personal protective equipment (PPE), standard operating procedures (SOPs), and administrative (leadership) controls. If we apply the containment philosophy only to the agent, we are missing the second greatest risk.

As mentioned before, we can spend millions of dollars on engineering controls, thousands of dollars on PPE, and hundreds of hours writing SOPs, yet in an instant the behavior of an individual can negate all of these controls. For this reason, the containment philosophy must be applied to staff working with agents and patients to control risks associated with human nature. Multiplying this risk by adding staff magnifies the situation.

A safety culture is a group of individuals within a specified organization who share common risks, adhere to clear rules, and demonstrate consistent rituals aimed at mitigating the risk as a collective group. The containment philosophy must be applied to agents, people, and organizations. Failure to apply the containment philosophy to all of these will leave a huge gap in safety programs.

THE AGENTS

Here, I discuss the philosophy of containment for each primary control of safety and how we practice containment. In almost all literature specific to biosafety, engineering, PPE, and SOPs are discussed before leadership controls. But most failures in containment arise from the lack of leadership controls within organizations, so let's begin there.

Leadership Controls

Regardless of whether we are in hospitals or laboratories, on the football field, or at home, safety programs suffer if leadership is not fully engaged. "Engaged" means that leaders support programs that mitigate overall risk to the health and safety of those working with patients and agents. Lack of leadership leads to lack of awareness specific to common risks. Lack of awareness of common risks leads to the lack of common rule development. Failure to have clear rules leads to the lack of consistent rituals among a group, which mitigates overall risk to them and the organization. I have seen safety cultures that have leaders, but I have never seen a safety culture exist without leadership, which equates to nothing more than clear expectations, accountability of individual behavior, and compliance with organizational rules. Leadership is not a position of authority but a philosophy practiced by any and all who choose to practice it. This is why cultures can exist without a single leader, relying instead on collective leadership within the group. The following are examples of leadership controls that must be implemented to ensure that the containment philosophy is applied to control the agents.

Surveillance programs. Leadership must be directly involved with occupational health, medical, and incident surveillance programs. Occupational health programs are different from medical surveillance programs. Occupational health programs typically provide access to a trained medical professional for a variety of needs. These medical providers can provide initial screening aimed at determining a "fit-for-duty" status. For instance, a person might be willing to work with biological risks, but the person's body is predisposed to increase overall risk to their own health and safety. The thought of any organization allowing individuals to work around infectious diseases without prior screening seems extremely irresponsible. Because human health changes frequently, the same medical providers can provide annual screening to ensure staff remain in a fit-for-duty state. If an exposure occurs, occupational health program staff can recommend treatment options and serve as advocates to other health care professionals when exploring treatment strategies. Even if an occupational health program exists, how and when do staff feed into the program?

This is where medical surveillance programs come into effect. Leadership cannot just support occupational health programs; they must also ensure that anyone who works with a biological agent or sick patient reports any symptoms that match the clinical presentation of the risk they are working around. Because most emerging infectious diseases present with influenza-like symptoms, staff might not consider the work they do to be the cause of the symptoms they are experiencing. Medical surveillance programs place a laboratory director in charge of receiving communication should a staff member begin experiencing symptoms that match the clinical presentation of the agent in question. If this person is the only one in the laboratory experiencing the symptoms, with no noted exposures, very little attention should be given to this situation. However, if multiple people within the same laboratory become ill within a specified incubation period of the biological risk they work around, all staff must be interviewed and asked to report any known incidents, accidents, or near misses. If none are identified, discussion about their personal interactions with the ill staff member could be explored; i.e., have they eaten or spent time together outside work? If the laboratory or clinical setting seems to be the only place their paths crossed, it makes sense to explore the workplace as the cause of the illness. Leadership involvement in medical surveillance programs increases overall compliance, which reduces risk if the hospital or laboratory environment is causing illness because it can be quickly discovered and mitigated before additional illnesses occur.

People might fail to report an accident or incident if, at the time, it did not seem to be an exposure or if they fear punishment from leadership. Only after the incubation period occurs do they realize that an exposure occurred, when they are ill and perhaps sicker than they would have been if they had made a report at the time. Trust in leadership produces incident surveillance programs that ensure that accidents, incidents, and near misses are reported, leading to safety improvement (as a result of lessons learned) and allowing staff to be treated with prophylactic treatment (if available), to be monitored throughout the incubation period, and to be isolated for the safety of others if they become contagious.

Failure of leadership involvement in surveillance programs leads to multiple risks around biological agents. Staff must be fit for duty when working around infectious diseases. Failures in medical surveillance programs may produce multiple illnesses from a common laboratory environment. Last, failures in incident surveillance could lead to infections that could have been treated prophylactically or that lead to secondary transmission as those who are ill continue to work with those who are not.

Training programs. As I write this chapter, it is summer at the Kaufman home, and my three children are around, reminding me just how important leadership is related to training programs. For example, if I gave the children an option to stay home and play video games or go to summer school, you can guess what they would choose. But if summer school is needed, the leader must make the decision and make that happen. Leaders don't ask whether their scientists would rather do science or attend a training program. Leadership must advocate safety and ensure that the appropriate training takes place. If my kids have special educational needs, I make sure the school I send them to has the teachers and resources to actually make a difference for my children.

If leaders are called to prepare the workforce, they need to ensure that good training is part of that preparedness. Leaders must make sure the training program accomplishes what they aim to accomplish: does it increase awareness, abilities, or application and comprehension of the content they teach? Leaders should not advocate just any kind of training but must care about the quality and frequency of training, too. Leaders must advocate for more than the typical 1% on-the-job training that most laboratory staff receive (if they are lucky). Unfortunately, much of safety training today aims to check regulatory boxes, leaving scientists with a negative attitude toward safety training in general, especially if there are "canned" programs that they are forced to attend over and over.

Accountability programs. Leaders are also called to protect the workforce. This is done mainly in two ways: by supporting those who follow the rules, even if mistakes are made, and by addressing those who choose to break the rules. Compliance will occur only if accountability to expectations is facilitated. It is within both positive accountability and negative accountability that compliance with a universal set of rules, developed to produce consistent rituals that mitigate common risks shared by all within an organization, occurs.

Accountability to universal rules demonstrates that leaders care, as in "It matters that you are doing what I am asking you to do." If leaders don't care whether someone follows rules or not, why would rules be followed? If rules are not followed, consistent rituals do not occur, thereby producing inconsistent safety results. Simple, frequent recognition by leaders that the workforce is following universal rules goes a long way when it comes to increasing overall compliance. Additionally, if a leader immediately addresses insubordinate behavior, in order to protect all those at risk, this sends a very strong message of caring to the workforce. This solidifies a culture of safety practices among those who have different backgrounds and experience and education levels. In short, accountability is the keystone of safety culture development, whereas leadership engagement is the sustainability of safety culture within organizations.

There can be leadership without safety, but there is no safety without leadership. Leadership is the main ingredient in a safety culture. When leadership shows that an organization cares about what you do, you can begin developing clear rules, as discussed below.

Standard Operating Procedures

My mentors taught me a very simple concept when it comes to the containment philosophy specific to SOPs. "Never mix clean with dirty. Always do your best to work from clean to dirty and dirty to clean. That is the golden rule for containment." You would be surprised to find that many programs do not follow this basic rule, and as a result mix clean and dirty. A common example of this can be found in most emergency rooms around the world. Why on earth would we seat febrile patients next to patients who have broken arms?

The following are examples of SOP controls that should be considered to ensure the overall containment philosophy within the organization.

Quantity. Most SOPs are written more to appease regulators and leadership than they are to teach the workforce how to behave. Seriously, I have seen SOPs for how to write SOPs and even some for how to use the bathroom. Can you imagine a leader holding someone accountable for the bathroom SOP? The term "SOPs" should mean something to all laboratory and health care staff. They are separate from policies and guidelines. An SOP is written because if you do not behave in a standardized fashion around a specific risk, health and safety are in jeopardy. This means we do not write SOPs for everything. In all the years I have been serving biosafety, I cannot think of more than 20 SOPs that need to be written for the sake of safety. I understand that science may need more SOPs, but safety does not.

Quality. The quality of SOPs should be measured by risk reduction, practicality, and workforce acceptance of the SOP. Effective SOPs are quality SOPs. The quality of an SOP is best measured through the satisfaction among the workforce for its ability to mitigate overall risk and keep them safe, not by regulator and leadership satisfaction. You can try to satisfy both groups, but what works for the workforce will not always work for leadership or regulators. The workforce needs, more or less, a checklist with behavioral cues; leaders and regulators need explanations and details.

Flexibility. When something is done for the sake of safety but actually increases overall risk, this is an example of "reverse safety." Unfortunately, SOPs can be a type of reverse safety if there is an inability to modify them in a quick and easy fashion. Organizations can put up barriers that prevent the ability to change existing SOPs. Remember that standard operating procedures are actually standard operating *behaviors*. Whenever we ask someone to behave in a repeated fashion, innovations and better ways of doing the behavior are developed.

Often, the SOPs I review are more than 10 years old. Think about what that means: for at least a decade, you have been doing this behavior, never learning a better way to do it, and without a single improvement. Consider whether your SOPs are preventing growth within your organization. If you are not changing, you are not growing. When organizations are flexible and allow changes in SOPs, that organization can mature and become better at mitigating the always-changing risks they face.

Identifying the most important rules for SOPs, ensuring that the SOPs are actually effective, and providing processes for improving SOPs together set the stage for providing resources that are needed to work safely with infectious diseases.

Personal Protective Equipment

The goal of PPE is very clear: to keep what we are working with where we are working with it inside the containment areas and, by protecting all portals of entry, to keep what we are working with outside our bodies. Unfortunately, the use of PPE can present perhaps the greatest risk that health care staff and laboratorians face throughout their careers. Consider the following to ensure the containment philosophy among staff working in PPE.

Comfort of PPE. I sometimes tell my children, "Suck it up, buttercup!" when they are complaining about certain things in life. However, I would never make this statement when health care or laboratory staff complain that the PPE they wear is uncomfortable. Discomfort will prevent proper use of PPE; I once heard a scientist say that he would not wear the PPE (N95 respirator and eye protection) because it blinds him. He was told, "It must not be fitting right. Make sure it fits right, and it won't do that." In order for containment to work, we must listen to complaints about PPE. If staff says that they don't trust the PPE or that it is uncomfortable, listen to and address these complaints immediately. Solutions can always be found. Failure to wear PPE while working around infectious disease agents can lead to breaches in overall containment.

Selection of PPE. Everything you do in biosafety should be driven by a risk assessment, including choosing PPE. If this is being done properly, why are people who work in BSL3 laboratories wearing full hoods? That is appropriate if you work in ABSL3 laboratories, which may house animals that fling feces or where splashes can occur in unpredictable fashions. But do you really need a hood to sit at a cabinet and culture?

I have been guilty of promoting this specific behavior, but that was when there were no other choices. I recommended hoods from MaxAir because that was all they had. They now have a cheaper and more practical product available that saves time and resources. When we choose PPE, we must make sure the selection is supported by the risk assessment but is also comfortable for those who will be asked to wear it. N95s may be cheaper than powered air-purifying respirators (PAPRs), but I have seen firsthand how negative-pressure respirators lead to greater levels of fatigue among laboratory and health care staff. Fatigue among the staff increases the risk of making mistakes. Select your PPE wisely, based on risk assessments *and* staff feedback.

Practicality of PPE. This is not a shameless plug for MaxAir PAPRs, but in my opinion they are the most practical PAPRs on the market for working in most laboratories and health care settings. They are worn under all protective gear, have small batteries, and can be easily decontaminated. To my mind, the competitors have yet to come up with a design that beats the practicality of the MaxAir system. That does not mean it will always be this way, however.

Listen to the concerns of your staff. If comfort and practicality are being questioned, be prepared to make a quick change in PPE to ensure compliance with SOPs. You can have the best SOPs, but if the PPE is not comfortable, selected by staff, and practical to use, containment will be challenged as a result of PPE failures.

Engineering Controls

When talking safety, engineering controls are usually the first aspect discussed. What risks can we engineer out? My perspective is that, as long as humans are involved, human behavior will always be the source of the greatest risks we face. However, engineering controls are extremely important within the context of the containment philosophy. Failure to integrate the engineering controls will certainly mean the loss of containment and hence an overall failure in the containment philosophy.

There is no doubt that engineering controls tend to be the most expensive of the four primary controls of biosafety. Directional airflow, high-efficiency particulate air (HEPA) filters, biosafety cabinets, and autoclaves are among the many controls that ensure sustained containment of biological agents. Engineering controls negate risks that deal specifically with air, liquids, and physical waste, all of which require PPE, effective SOPs, and training to be used properly. The minute a human interacts with engineering controls, however, the risk of failure among these controls begins to increase.

THE PEOPLE

Until robots replace those working on the frontlines of emerging infectious diseases, human risk factors will remain a critical issue in the application of the containment philosophy. I wish we could program humans to behave in specific ways at specific moments; this would help us successfully respond to risks. One day, we may be able to program people to be safer drivers and avoid unnecessary risks, etc., but for now we must live with the reality that human behavior poses direct risks to overall containment. There are three top human risk factor issues that directly impact effective containment of biological risks: basic needs, complacency, and being human.

Basic Needs

It would be nice to think that people come to work because they are loyal to the organization they work for, but people tend to work because they have needs of their own. These basic human needs pose a substantial risk to the containment philosophy. When someone puts his or her needs above the collective needs of others, containment is at risk. There is no better example today than cell phone usage among health care staff and laboratorians in areas where and at times when they should not be using them. Cell phones are used to satisfy basic needs on multiple levels for these individuals, even though they know that it is not the safest or most responsible behavior. We also see this when drivers use their cell phones. People will agree that it is an unsafe source of distraction, but then text and drive, demonstrating a direct conflict between what they think and what they actually do.

Let's say that I am a scientist who has three children attending college. I have huge financial responsibilities to those I love more than myself. If I make a mistake at work, I fear losing my job as a result of reporting this mistake. What do I choose? Some reading this would see this as an issue of integrity and character, but as a behaviorist, I see that these are issues and labels of cognition. Simply stated, people will (usually) behave in their own self-interest first. Of course, there are outliers who place others before themselves, a very rare commodity.

Humans will take care of their own needs. If the organization provides a paycheck to its workforce and nothing else, then the job is only a means to pay bills and provide for the family. What if the workplace provides a sense of belonging as an important component in the workplace? To belong matters most, more than being paid. This is a simple strategy that can make someone an important part of the organization. Then they are a part of something greater than just themselves.

What does it take to belong? To apply the containment philosophy within the organization, it takes common risks, clear rules, and consistent rituals.

Complacency

I like facilitating individual interviews when serving clients. It helps me differentiate safety climate—what they say they do—from safety culture—what they really do. During this process, I provide a list of 10 interview questions. Those I am interviewing are asked to pick three questions specific to safety. When they have answered their three questions, I ask them to choose a question that they would like me to answer. I have done this exercise at several locations with hundreds of professions, and the number one question they choose for me to answer is, "How do you deal with complacency in the workplace?" A couple of years ago, I began asking each of them whether they were experiencing this personally. Almost everyone answered yes.

Humans are designed to become complacent. Our brains are bigger, better, and faster than we can fully realize. When we learn new behaviors, our brains are fully engaged, absorbing details, looking for patterns, and even assessing overall risk. We absorb so much and then, within short periods of time, our brains solve the "equation" of our situation, develop habits, and go back to sleep. This concept is supported by neuroscience and allows us to multitask. We can actually think, solve problems, and even participate in multiple behaviors while doing a common or frequent task. The intriguing part about complacency is that most human beings can admit it! They may call it boredom or give it some other label, but the fact is that it becomes more difficult to pay attention to specific details when a behavior becomes habitual and uneventful.

Containment is at risk when humans are no longer paying attention to simple details like glove removal, patient triage, working in a biosafety cabinet, and management of waste. Habits take over, multitasking and thinking about other things begin, and details that were once seen are completely overlooked, not intentionally but because we are human.

Being Human

This brings us to being human. Every day I go to work I wish I were getting younger and my immune system getting stronger. I also wish that I would never get sick, never stress, and never think of things in life that are far more important than the risks I encounter within a laboratory or hospital. I do wish to see my kids graduate, have grandkids one day, and experience all the positive things in life. I will also live to see tragedies, lose loved ones, and endure other losses that may forever change my life. I am human, and life happens to me, not just around me.

Because life happens to you, containment is at risk. Some days you come to work ready to work, full of love for what you do and who you are working with. At other times, it's the opposite. Sometimes life is good, and sometimes it is not. These things change perceptions and attitudes, which directly affect overall behavior and therefore overall safety and containment.

The containment philosophy can only be applied if we recognize what we must contain. Unfortunately, much focus in biosafety has fallen on the agents themselves and less on the people or the organization working with the agent. Understanding human limitations is critical for maintaining the containment philosophy. Give people PPE, and write as many SOPs as you want, but if you fail to understand staff through the lens of being human, there will be an enormous gap within all existing safety programs.

THE ORGANIZATION

When I lead training programs internationally, I usually use the term "family" to discuss safety culture, because most of us can relate to the concept of belonging to a family. A family is an organization that subscribes to a unique set of characteristics, which bind it together and help increase the overall chance of survival. Should we not be doing the same thing, creating a sense of belonging, in hospitals and laboratories?

The containment philosophy within organizations is solidified when the following three items are integrated throughout the organization: the biological agent, the individuals dealing with it, and the culture of safety within the organization as a whole. Together, a collective effort needs to be applied that minimizes risks not solely to individuals but to the reputation of the organization, loved ones, and even the public.

The containment philosophy begins with the biological agent. It then passes through the people working with the agent. It finds a home within the culture of the organization. All of the individuals within the organization share a set of common risks. The first and most obvious risks are to health and safety; however, risk also extends to the sustainability of employment. If containment is lost, the reputation of the organization may also be lost, leading to the loss of jobs.

Additionally, the public tends to be quite hard on science when containment is lost specific to a research laboratory. Hospitals face heavy public scrutiny anyway for treating patients who are sick with emerging infectious diseases. If there is a loss of containment, the implications can be catastrophic. Public outrage damages science and quality health care. Regardless of whether you are a cleaner, laboratory technician, principal investigator, doctor, nurse, CEO, or vice president of research, failure in containment is a risk that everyone in the organization has in common.

Common risks are the main driver of common rules. It does not matter who you are or where you sit in the organization; all rules apply to all levels within the organization. The rules are clear, to ensure that expectations are communicated to all who belong in the organization.

Belongingness in the organization is the acceptance of both the common risks and clear rules that aim to mitigate the overall risks assumed by all. The containment philosophy includes an attitude of unity—family, team—not one where individuals matter more than the collective group.

When common risks are identified, and clear rules are established, consistent rituals (behaviors) begin to occur. It is a beautiful thing to witness. We see this among firefighters, police officers, and military personnel and throughout the aviation industry. Different people, with different backgrounds and different experiences, are behaving as one, working under a set of clear rules for the sake of mitigating common risks.

The institutional culture of safety remains the greatest risk when working with emerging infectious diseases. There is no doubt that the biological agent itself is a risk. However, years of mitigating risks associated with infectious diseases provide insights and methods that have been proven effective and stand the test of time. We also know that the people working with the agent pose a threat, because being human has inherent risks that produce unavoidable limitations. The containment philosophy controls risks associated with both the agent and people who work with the agent. However, how an organization chooses to implement the philosophy of containment will form the culture of safety within that organization.

Common risks associated with the agent are shared throughout the organization. Clear rules are written for those in the organization working with the agent. Consistent rituals create the collective behavior of the organization, which drives the collective beliefs of individuals, which in turn form the safety culture of an organization.

CONCLUSION

The goal of this chapter is to discuss the overall containment philosophy specific to the agent, people working with the agent, and the culture of the organization. There is no shortage of information for controlling the agent. There is no shortage of engineering, PPE, and plans to mitigate risks associated with the agent. However, when we begin exploring controls for dealing with the people working with the agent, there is not much to draw from. It's even harder to train for how collective behavior impacts overall safety. My point is that if the containment philosophy is the action of keeping something harmful under control or within limits, we must look well beyond the agent. We must equally address the individual and the group of individuals (culture) to adequately keep something harmful under control and within containment.

BIOSAFETY *in the First Person*

I'd Do It All Over Again by Lee Alderman

Lee Alderman is like a second father to me. The hours that Lee has spent teaching me about biosafety, life, and leadership will never be forgotten. Lee is a man of great integrity. He believes that biosafety is about service—never obstructing the progress of science but ensuring that the work of science is done safely. His mentor was John H. Richardson, for whom the Association for Biosafety and Biosecurity (ABSA) International John H. Richardson Award is named. In his turn, Lee has been my greatest mentor, and I owe much of what I have accomplished in my career to him. It is with respect and gratitude that I present, in his own words, Lee Alderman's story of his career in biosafety.

My biosafety journey began at the CDC Technical Development Laboratory (CDC/TDL) facility in Savannah, Georgia, in 1962, where I was hired as a chemist laboratory technician and eventually worked as an analytical chemist. Looking back on my years at the CDC, I can recall events that had a significant impact on the way I approached my work in the laboratory setting.

An example of this occurred in the mid–1960s. I had been working in the CDC/TDL as a chemistry technician for about three years and was at an age when you think you are invincible. On this day, I had just prepared a solution of sulfuric acid that was required in a colorimetric analytical procedure to determine concentration of pesticides in air samples. The procedure called for the addition of 10 ml of the diluted sulfuric acid solution I had prepared. This required me to draw the acid solution from a 1-liter volumetric flask using a pipette. I had done this many times by mouth pipetting without accident. On this day, my laboratory colleague Brenda Jenkins walked into the lab area to begin her work just as I started to mouth pipette the sulfuric acid solution from the flask. I was startled by her entry, and I accidentally got the acid solution in my mouth. Instinctively, I released the pipette and tried to spit the acid from my mouth; at the same time, the pipette dropped into the flask, which broke, and the freed sulfuric acid solution wetted me from the waist down. I could feel the acid getting warm on my skin. In the lab, the sinks were large and reasonably low to the floor, so I climbed into one and turned the water on to dilute the acid and cool my skin. I suffered no ill effects, except for being embarrassed.

What did I learn from this? In school you are taught not to mouth pipette, but like most of my colleagues most of the time, I did anyway. However, for the remainder of my laboratory career, I used a rubber bulb or a mechanical pipette device when pipetting, and whenever I recalled this incident, it reinforced my using a mechanical pipette device.

During those early years at the CDC/TDL, we were trying to develop new pesticides that would replace DDT, which was being banned, and that could be used to control mosquitoes carrying malaria, which was causing hundreds of thousands of people to die annually. The analytical technique we used in those early years was a colorimetric determination method with sensitivity in the microgram range. Over the next five years, new analytical instrumentation was developed that increased the sensitivity from microgram to nanogram detection levels, which is a thousand times more sensitive. During the next 10 years, analytical instrumentation improved so that it could detect levels in the picogram range, which is another thousand times more sensitive.

In 1972, the CDC/TDL facility closed, and I was transferred to the CDC in Atlanta, Georgia. In 1977, I had the opportunity to leave my laboratory position and move into the biosafety office program as an assistant biosafety officer supporting the operation of the first BSL4 laboratory at CDC. My job was a facilities position, meant to ensure that the laboratory mechanical systems and equipment were working as designed. My responsibilities included developing an operational checklist that was performed every morning before the laboratory staff entered the laboratory, decontamination of the laboratory so that supply and exhaust air HEPA filters could be tested, certification of biological safety cabinets in the laboratory, and coordinating required maintenance during annual shutdown. Over the next 10 years, CDC realized the need for additional laboratory space and a larger BSL4 facility. Using our experiences gained from the first BSL4 laboratory, we worked with the laboratory staff, architects, and engineers to design and construct a new BSL4 facility that was completed in the late 1980s.

In 1992, I retired from CDC after 30 years to accept a position at Emory University School of Medicine as the university biosafety officer. My responsibilities included not only the School of Medicine but the Yerkes Primate Center (YPC) as well. The YPC is located near the Emory main campus.

I recall a tragic event at the YPC that altered the way I functioned as a biosafety officer. I will never forget the call I received in 1997 from Harold McClure, senior research veterinarian at YPC. He said, "Lee, we have a problem." He went on to tell me that a young female employee, Beth Griffin, had been exposed to herpes B virus, had become infected, and was in the hospital. The exposure occurred while she was moving a caged rhesus monkey from one location to another. It appeared that the monkey had flung some waste from the bottom of the cage at Beth, some of which entered her eye. Several days later, she became symptomatic for herpes B virus infection and was admitted to the hospital. At first, she responded well to the treatment, but several days later she developed a fever and paralysis, and she died about five weeks later. This event was a sad occasion for all the YPC and Emory staff.

Since 1933, there have been 40 cases of human-to-primate transmissions reported, almost all the result of bites and scratches. On the day Beth was infected, she was wearing a lab coat, a mask, and gloves. The cage was covered with a fine wire mesh to protect her and others from bites and scratches. Because of this tragic event, Yerkes modified the safety clothing to require eye protection. In addition, a full-time safety officer was added to the Yerkes staff.

The impact on me personally was sadness that this had happened to the bright 22-year-old woman and to her family and friends. As I said, this changed the way I worked. I decided that I should frequently visit the laboratory staff, making myself available and trying to identify potential problems and providing solutions when I could. I feel that, to be the most effective biosafety officer, you must be seen in the lab, you must be trusted, and you must do what you say you will do, when you say you will do it.

In 2001, I was selected as the director of the Environmental Health and Safety Office, with a staff of some 40 members and the responsibility for biosafety, chemical safety, radiation safety, and industrial hygiene for Emory University.

In 2006, I retired from Emory after 13 years and started my own biosafety consulting business. In this capacity, I continued working with Sean Kaufman, presenting biosafety training classes in the United States and many countries around the world. I was fortunate enough to find my working niche, and I can also say that, given the opportunity, I would do it all over again.

chapter 8

Plans + Behaviors = Outcomes

Many of us want to lose weight. We develop a plan that includes increases in activity and decreases in calorie intake. We identify an outcome and promise ourselves that we will lose 15 pounds in the next three months. If I were hired as a consultant, I would tell you that the plan is sound and so is the outcome. If you burn more calories than you take in, you will lose weight. Your outcome is clearly identified, as you have quantified the overall weight loss and identified the amount of time in which you plan to lose the weight. Your plan and outcome look great. However, three months later, you weigh more than you did before and you failed to achieve your overall outcome. That's because you were missing something important.

Several years ago, I came up with the concept that plans plus behaviors equal outcomes ($P + B = O$). In addition to a plan, you need to account for behavior. A behavior alone produces an outcome without a plan ($B = O - P$). Typically, we behave without a plan when we aren't challenged by a potential crisis or the lack of a basic need. This kind of behavior produces an outcome that usually cannot be predicted, because the behavior occurred in response to an environmental stimulus rather than being a planned behavior.

A plan alone produces an outcome minus behavior ($P = O - B$). This is a major fact to remember. Many failures—personally, professionally, even within humanity—occur because too much faith is put into a plan that is supposed to deliver a desired outcome. The problem is that when a human begins to respond to the plan (behave), the outcome can change as a result of the variety of behaviors among different people. Additionally, although behaviors can be planned, environmental influences, perceptions, and attitudes can have a direct influence on behavior, all of which affect the overall outcome. In short, a plan is simply an outcome minus human behavior.

Our company slogan is "Let's do safety together." It recognizes the importance of behavior as the bridge between an effective plan and a desired outcome. Behavior is and should always remain the keystone of any plan and desired outcome. If you desire safety, you can write all the plans you want and dream of outcomes that set the safety gold standard, but if you do not focus on how staff are behaving, there will be no standard of safety. Instead, you will get a variety of safety outcomes that could be good *and* bad.

EFFECTIVE PLANS

A good plan is one that aids in producing the desired outcome. A plan alone has never produced consistent behaviors and outcomes and never will. Much more is needed to achieve the desired outcome.

An example of this is wearing protective gloves during laboratory work or when working with patients. During the early stages of the HIV pandemic, doctors, nurses, and dentists learned that treating patients with bare hands, which might have cuts and tears, especially around the finger cuticles, placed them at increased risk for exposure to blood-borne pathogens. The plan was to provide gloves to minimize this risk: a good plan with a good desired outcome. However, watch the variety of ways that individuals, including health care staff and laboratorians globally, remove their gloves, and it should concern you. During trainings, I have witnessed thousands remove gloves that had been contaminated with GloGerm, a substance that glows under a black light. It's pretty typical to see, after glove removal, GloGerm in and around the cuticles, usually on the index finger. Why? Well, most glove removal plans are ineffective. Removing gloves correctly is as important as wearing them in the first place. Of all the behaviors a scientist or health care worker will do, glove removal will have the most impact on safety. Let's do the math:

Average glove usage per day = 20 pairs of gloves

20 pairs = 40 gloves per day

40×5 days a week = 200 gloves per week

200×48 weeks a year = 9,600 gloves per year

$9,600 \times 20$ years of work = 192,000 gloves in 20 years of work

$192,000 \times 0.01$ contamination factor = 1,920 gloves contaminated

$1,920/2 = 960$ times in a 20-year career where proper glove removal will matter, assuming a 1% contamination rate

These numbers are conservative. The fact remains that if you are working with agents or patients that could make you sick, you need proper personal protective equipment as part of an effective plan plus effective behaviors to properly address your risk. You cannot have effective behaviors if your overall plan is flawed.

I use the acronym PLANS to remember what makes up effective plans. Effective plans are practical, learnable, applicable, necessary, and specific. Using glove removal as an example, I will discuss each of these aspects.

Practical

A plan is not a theory; it is an applied concept. So, when you are developing plans, it is important to envision your goal, but you must make sure the plan can be achieved by those following it. Theories are something people talk about. Plans are something people do. An effective plan must be proven possible to perform; otherwise, it is nothing more than an unproven concept.

Learnable

Those reading the plan must be capable of learning it. I have witnessed many plans that are, to be blunt, not learnable. Maybe it is a language issue. Sometimes it is attitude issue, e.g., there is no need to learn. Most of the time it is resource issue. For example, you can't learn how to swim if there is no swimming pool available. Plans must be able to be learned.

Applicable

Nobody can support a plan that has no relevance and may not even be appropriate. It is seen as wasteful. Unfortunately, many plans given to laboratorians and health care staff cause frustration because they do not apply to their daily activities. Plans must be relevant and have an obvious purpose to be effective.

Necessary

If nobody cares whether a plan is being followed, it is just words on paper. Plans must be required, with a differentiation between those who follow plans and those who do not. Science has already demonstrated the vast increases in the success of losing weight when you do it with a friend. Friends hold each other accountable to the plans they develop together to lose weight. If there are no consequences to not following the plan, meaning it is not required, then the plan does not work. Plans must be necessary to be effective.

Specific

I like to say a plan is simple, but this may not be the case. Plans are specific in the sense that they clearly identify expectations. These expectations can be both behavioral and cognitive expectations, but the more specific these expectations are, the more effective the plan will be. (For more on this, refer to the internal validity section of chapter 12). Plans must be specific and provide clear expectations to be effective.

Those who write plans usually do not spend time thinking about whether a plan is practical and learnable. They may not ask themselves whether the plan applies to the work being done or if it is required for the work to be completed. Even if plans have all of this, they can lack specificity, making them less effective.

SUSTAINED BEHAVIORS

A plan requires sustained behavior. A sustained behavior is one that is practiced not just once but several times, in a specific place, over an extended period. One of my favorite health education theories is the health belief model, which inspired me to identify the five things I believe are needed for sustained behavioral practices. These items are (i) understanding of risk, (ii) understanding of benefits, (iii) access to resources, (iv) abilities, and (v) self-efficacy. Miss one of these important factors and sustained behavior does not occur.

Understanding of Risk

In order for an individual to behave a certain way, there must be a reason for it. One good reason is to avoid risk. How one defines risk can make an enormous difference in behavioral compliance. For example, if risk is defined in terms of individual health and safety, this will be a problem, because each individual can define the risk. When individuals begin defining risk for themselves, they also begin choosing behaviors that match their overall understanding (perceptions) of the risks they face.

If, however, the risk becomes the inability to perform their duties—for example, if there is an accident, the company will be closed—then it is compliance with the behavior (or plan) that determines overall risk. Perceptions and attitudes no longer matter because the risk is not one for the individual but for the collective group. This is why safety cultures are built on compliance and accountability. A safety culture shares common risks, rules, and rituals. It is these three things that bind different people with different backgrounds, experience, and education levels together. In short, people must understand what risk they face if they choose not to behave.

Understanding of Benefit

Once risk is defined, the discussion of what is gained by doing the behavior is needed. It is easy to motivate someone to behave when they are scared of something. However, fear is only temporary, and the more experience (especially if it is successful) someone has around a risk, the less scary it becomes. This loss of respect around the risk is the reason that gun and motorcycle injuries occur to some of those most experienced. Again, the benefit of risk avoidance should not be from a personal safety and health standpoint, it should be to allow you to remain part of the workforce, continue to do your job, and contribute to serving patients or biological sciences. The motivation to behave in a given way (risk and benefit) cannot be up to an individual but must be a collective effort. This can be done by shifting the risk-benefit equation away from the most obvious risk and to one of belongingness within an organization, team, or family.

Access to Resources

If we ask people to exhibit certain behavior but fail to provide the resources needed to comply with the behavioral request, that behavior will not occur. If the behavior does not occur, the plan does not work, and the desired outcome is not achieved. Plans require resources, and these resources must not be temporary but must be as sustainable as our plans are. I've seen people who lack this foresight invest significant time writing plans that burn through tremendous resources quickly, making the resources unavailable. So, these fail, too. You cannot lose weight if your plan is to go running after work and all you have is high heels.

Abilities

Lack of ability to behave is the greatest valley of sustained behavioral practices. This is where mentoring in both hospitals and laboratories is important for two reasons. First, mentoring teaches behavior. Second, mentoring verifies that the competency of an individual meets an identified standard. I hope this is included in your organizational plan.

Self-Efficacy

It takes some people more time than others to learn behaviors. One of the greatest differences between people is their levels of confidence. Some people have been raised to take chances, try the unknown, and believe they can do things without really being sure, whereas others grew up in families where taking chances and boasting that you could do something you had never tried before were discouraged or even punished. Self-efficacy, one's belief that he or she can successfully do the expected behavior, is important. Even if people understand the risk and benefits, have access to resources, and have demonstrated the ability, lacking the belief that they can successfully carry out the behavior when the time comes inhibits sustained behavioral practices. Self-efficacy is gained through repetition, and some need more repetition than others. Self-efficacy is gained by asking someone to do the same thing over and over again until they say, "I got this."

Behavior is the bridge between plans and outcomes. Any person who is developing a safety plan, whether for laboratories or for hospitals, must ensure that the staff following this plan have access to all five items needed for sustained behavior to occur. When I am asked to determine why behavior is not occurring according to a specific plan, 99% of the time it is because one of these items is clearly missing. This is easy to spot when interviewing the one who is not behaving according to the plan. The next time there is someone on staff who is not behaving according to an effective plan, here is the interview you need to conduct.

1. *Do you understand the risk?* They will say "yes," so ask them to explain the risk. When they explain the risk, remind them that the risk affects not only them but also those they work with, their family members, and the reputation of the organization.
2. *Do you understand the benefits?* They will say "yes" and will have learned from your lesson above to include benefits to those beyond themselves.
3. *Do you have access to resources needed?* If they say "yes," ask them what resources are needed to follow the plan. This can cognitively verify whether they know the plan, as knowing the plan requires knowing what is needed to behave according to the plan.

4. *Do you have the skills needed?* If they say "yes," ask them to demon-strate the behavior for you right then. This is behavioral verification.

5. *Do you believe you can successfully do this plan when we need you to do it?* Ask them when they would *not* be able to do the behavior.

You will learn much from this interview; undoubtedly you will learn something unexpected. When plans are developed, we fail to think about what is needed to obtain the overall outcome of the plan. When we are cognizant of those needs, the likelihood of an effective plan producing a desired outcome is much higher.

DESIRED OUTCOMES

One of the most frustrating aspects of serving biosafety is that the outcome of biosafety is not clearly defined. The same holds true for the outcome of infection control. How do we know if a biosafety or infection control program is successful? Is it zero secondary infections? Is it a decrease in accidents and incidents? One of the biggest weaknesses in both the professions of biosafety and infection control is the lack of defined outcomes. We need to know what we are looking for when these programs are implemented.

Desired outcomes are both measurable and time specific. This means that to have a desired outcome, you must have a baseline. Can anyone tell me how many laboratory-acquired infections we have? How about hospital-acquired infections? Naturally, we would look to the reporting of incidents and accidents, and the lower the number of reports, the better the safety programs, right? Wrong. If the reported numbers of incidents and accidents are decreasing, I suggest that what is happening is more hiding of true behaviors rather than reporting them. I would argue that good biosafety programs produce transparency unlike anything most of us are comfortable with.

We struggle in attempting to measure outcomes of biosafety and infection control programs. This indicates that both areas are in need of vast amounts of research so that we can establish desired outcomes. We should also consider the time it takes to begin seeing results as an indicator of implementation of new behaviors needed to achieve the desired outcomes. Still, it's difficult to set outcomes. What risk is truly acceptable for laboratory research? How many deaths are acceptable among health care providers when treating patients who are sick with emerging infectious diseases? These answers are key factors in determining our desired outcomes.

I will never forget serving the nurses and doctors at Emory University Healthcare during the treatment of the first two cases of Ebola in the United States. I assisted with the development of the plans and managed the isolation unit, ensuring consistent behavioral practices among those working around the patients. I went into this important work, going into isolation, with a very clear outcome in mind, partially developed by my wife, Jacqueline.

You see, because of the potential risk, I asked her permission do this work. Jacqueline said, "They have to do this your way, Sean, for two reasons." I asked what those were. She said, "If someone gets sick, you will never forgive yourself. So do not compromise. Make them do what you need them to do, because if they fail, you will own their failure." I asked about the second reason. Jacqueline continued, "You are a husband and father of three, Sean. There is risk in this for you, too. Come home safely, please." She helped make the desired outcomes clear to me. First, I knew both patients would either survive or not. I knew the amount of time I was going to serve. I also knew that success for me had to be twofold: (i) no nurses or doctors would become ill, and (ii) I did not become ill.

I cannot tell you how proud I was watching a CNN report about Emory while I was in Belgium and seeing Kent Brantly, the first person with Ebola virus treated in the United States, leave the hospital. All the nurses and doctors who had cared for him were behind him, and even though I never treated Dr. Brantly, I served those nurses and doctors. At first, I thought it was nothing—anyone could do it. But months later, in Texas, two nurses got sick treating an Ebola-infected patient. It was at that moment that I realized that what had been achieved at Emory was very special, a most desired outcome.

PUTTING IT ALL TOGETHER

It starts with a desired outcome, one that can be measurable within a specific period of time. Effective plans are written to achieve the desired outcomes, plans that are practical, learnable, applicable, necessary, and specific. However, it is sustained behavioral practices of those executing the plans that determine whether the desired outcome is achieved.

Behavior is the bridge between effective plans and desired outcomes. It has five pillars: understanding risk, benefit, resources, skills, and self-efficacy. If any of those pillars is missing, the bridge collapses. You can have effective plans and desired outcomes, but without behavior, all you have is hopes and goals on paper. It takes more than paper to achieve safety today, tomorrow, and in the years to come.

BIOSAFETY *in the First Person*

Biosafety Found Me by Anthony (AJ) Troiano

I first met AJ Troiano at a leadership conference in Atlanta. I have always been impressed by his no-nonsense approach and his desire to make biosafety a more respected and science-driven profession. I predict that AJ will do amazing things, and I appreciate his sharing a very personal story about his road to biosafety.

When the proverbial question "What do you aspire to be when you grow up?" is asked, the term "biosafety professional" most likely isn't among the answers. Throughout my undergraduate and graduate education, biosafety as a career path wasn't even close to being on my radar, mainly because I had no idea what biosafety even was. To the same effect, I've heard anecdote after anecdote from senior leaders in this field, admitting that they themselves stumbled into biosafety by happenstance. Personally, I wanted to be a physician, specifically an infectious disease physician, because I find microbes to be such fascinating organisms. Similar to how we can list the events or factors that precede a laboratory-acquired infection, in my life there is one incident that I can link to where my professional career path began.

On May 15, 2007, I had returned home that afternoon to Cape Cod, Massachusetts, from my freshman year of college at the University of New Hampshire. After a late night out with some of my friends who had also just returned home, I was awoken by the sound of crying, the flashing of blue and red lights, and the voice of a person I had never met before. It was a Massachusetts State police officer, who informed me that my father, Anthony J. Troiano, Sr., was clinging to life and being rushed to Rhode Island Hospital by ambulance.

At 19 years old, it's almost impossible to process the gravity of a situation like this. I just assumed there wasn't much risk involved in my father's long work days as a town manager. I certainly thought nothing was inherently risky in his 90-minute commute each way, or his severe sleep apnea, which medical experts argued most likely caused the car accident. He was only a few miles from home when the head-on collision happened. The driver of the other car was pronounced dead at the scene.

I watched my father get carted away into surgery that night at the hospital, with the expectation that he would die. Like myself, my father is an imposing figure at 6 feet 3 inches; his height is the reason why doctors were able to put a stent in his aorta that saved his life. It took nearly five weeks in a medically induced coma, several more in the ICU, and a stay at a rehabilitation hospital before my father returned home in August. Returning to school in September 2007 for the start of my sophomore year, I really didn't have much to say when asked, "So how was your summer?" by my friends and professors.

The fallout from the car accident was worse than the event itself, as during my fall semester, the legal case against my father began, and my GPA suffered tremendously from the stress that carried with it. As an honors program student, I needed to maintain at minimum a 3.2 GPA to keep my academic scholarships. It seemed so inconsequential to worry about grades as my father was convicted of vehicular homicide due to negligence, and I was powerless to do anything about it. I bore the name of a murderer, or so I was made to feel by the Internet and news media. School seemed like a wasted endeavor, and the last thing I wanted was to spend any more time in a hospital interning or shadowing physicians.

Organic chemistry? Failed. Ancient Greek? Clinging to a B. Biostatistics? Slept through my final. Premedical advisory meetings? Missed them all.

By my not passing organic chemistry, the predetermined "schedule" all premedical students stuck to like clockwork for classes and MCATs was ruined for me. Being a double major in microbiology and classics didn't help, either; I constantly had scheduling issues with all the laboratory sections of my science courses. The downward spiral continued into freefall, as my father lost his job in part due to his conviction, and our family was in the process of declaring bankruptcy not long after. How could I go to medical school if we were bankrupt? In retrospect, I was lucky to even go back to UNH for another semester!

Fast-forwarding to my senior year, I ended up working in a laboratory that studied microbial evolution of pathogens and found a niche for myself in scientific research. With medical school on the back burner, I decided to pursue a Ph.D. and stay in school for the foreseeable future. The job market wasn't very tantalizing in the spring of 2010, in part because of the recession that occurred after the subprime mortgage crisis of 2008–2009.

Luckily, I managed to find my way into the University of Connecticut Health Center's graduate school program in biomedical science. The reality is that I was waitlisted in the program initially, and only two days before the acceptance deadline I was offered a spot. Once there, however, I won the lottery when I found myself working under one of the most brilliant mentors I've ever had the privilege of knowing, Peter Setlow. Dr. Setlow did his postdoctoral work in the laboratory of the Nobel Laureate Arthur Kornberg, who first discovered the enzyme DNA polymerase I. Our laboratory studied bacterial spores, mainly *Bacillus subtilis*, but we collaborated on projects ranging from *Bacillus anthracis* to *Clostridium* (now *Clostridioides*) *difficile*. One of the most important lessons Peter taught me is that, as a scientist, we ask questions; however, the answer we seek is only sometimes "yes." More often than not, that answer is "no," but "no" is still an answer.

Graduating in the spring of 2015 from the University of Connecticut, I decided I didn't want to pursue a traditional career path in academic science and postdoc somewhere. The idea of chaining myself to a laboratory bench and writing grants until I was 75 seemed like insanity. I prided myself on my interpersonal skills, and I enjoyed interacting with other scientists at journal clubs and presentations, so I yearned for something more than what the bench could offer. So rather than jump headlong into a career like most of my peers, I returned to Cape Cod for a summer to decompress. It was the best decision I've ever made. For the first time in my young adult life, I had no responsibilities and no job prospects on the horizon, and I was able to spend time with my father in a way that was lost in 2007. My mother wasn't exactly thrilled that her 27-year-old Ph.D. son was moving back home, although I did enjoy her cooking again. I even went back to my old high school and college summer job as a lifeguard for two months at Craigville

Beach in Centerville, Massachusetts. In my mind, my friends struggling in their single-minded career path to find postdoctoral appointments were fools.

Living the quintessential dream came to a close quickly, as I began my career in public health at the Rhode Island State Health Laboratories in fall 2015. It was purely coincidental that I applied for what I thought was a clinical microbiology position, when in reality it was for their biosafety officer. The generic job description on the application didn't offer much information other than a two-sentence supplement at the end detailing that it was not a traditional clinical scientist position. I honestly didn't think much of it, because how difficult could "biosafety" really be?

It was baptism by fire on steroids, or perhaps baptism by napalm. Here I was with my doctorate, working in an area in which I had *zero* expertise. I used every resource at my disposal, including a generous training budget built into the grant that funded my position. My mission was to continue learning and grow into the biosafety field, using my scientific background to bridge the gaps between bench work and regulations. It helps tremendously to empathize with scientists, having been at the bench for several years. Sean Kaufman's lunch break series trainings were an invaluable resource (mainly because they were free), and this led me to his leadership summit in February of 2016.

At that summit, surrounded by people I didn't know, I did very little talking (which can be very difficult for me) and instead I listened. I absorbed every minute detail that the leaders of the biosafety world presented, and in particular, how they all got their start in the field. Every story was unique, and as I previously stated, it involved a lot of being in the right place at the right time rather than a dedicated career path. More important, it was there that I met Nina Pham and heard her story. From then on, my perspective of my line of work changed drastically, and I realized that there is a desperate need for leadership to emerge in biological safety. We stand on the precipice of newly emerging infectious diseases and scientific technologies that dictate it.

I now work in a field that, a few years ago, I didn't know existed and in which I received no formal educational and had zero colleagues to associate with. My path that led me here included a car accident, bankruptcy, a last-minute acceptance to graduate school, lifeguarding, and vague job application language. I guess my anecdote falls right in line with the rest of the community I've grown to deeply enjoy and appreciate. It's hard to understand perspective until after the fact, but I understand now that biosafety is exactly where I am meant to be.

chapter 9

Intrinsic Safety

What is your biggest failure professionally? Mine has to do with the concept of intrinsic safety. In 2004, I was asked by Ruth Berkelman to direct a program aimed at preparing individuals to work in high-containment (biosafety level 3 [BSL3] and BSL4) laboratories. At that time, there was a rapid expansion of high-containment laboratories, and there was a great need for a training program that would teach the skills needed to work safely in these laboratories.

Working with some very special people, I spent 10 years training workers at Emory University, throughout the United States, and around the world. In 2014, the National Institutes of Health grant ended. Ten years of work, and what was the result? Several hundred trained workers, granted, but nothing that would build on this limited success. The problem was that I myself was the training program. When the funding ended, that was it. What I should have been doing all that time was to build up a permanent training program that would have an impact on those working in laboratories and health care facilities around the world.

I admit I enjoyed that time. I had always dreamed of traveling the world, seeing places, and training people who would value what they were learning. When you travel to foreign countries where resources are limited, the hunger for information and desire for learning new behavior are unmatched. But rather than simply enjoying the experience, I should have focused on building the capacity in others to not only practice safer behaviors but train others to do so as well.

In 2012, when I knew the grant was running out, I started a small consulting company outside of Emory University. Now, when I was at the Emory training center, I could control the environment in my training sessions. When someone controls the environment, they can motivate you to do just about anything (if it is in line with your moral compass). But what happens when the environment cannot be controlled? People must make choices on their own. This was hard for me, as a behaviorist, to recognize, because it required an acknowledgment of the power of cognition and its influence on behavior. If I wasn't running the training course and could not control the environment, people would have to choose safety for themselves. What could I do to motivate people differently? It was then that I began the process to become a Certified Intrinsic Coach. I owe much of what I write here to the instructors at Intrinsic Solutions International. I have taken the lessons they taught me and applied them to training staff in biosafety and health care industries.

As fascinated as I am by how people behave around infectious diseases, I am also fascinated by how people are motivated to behave. There are three types of motivating forces: extrinsic, systemic, and intrinsic. Each is important and can be seen throughout organizations today. As you read about each of these motivating forces, see if you can identify specific examples of each in your organization. You may be surprised at what you discover.

EXTRINSIC MOTIVATION

There is no doubt that extrinsic motivation works. Unfortunately, it is usually very expensive and impractical. In 2014, I watched over the nurses and doctors treating Ebola patients. Within the isolation unit at Emory Hospital, any time the health care staff entered, exited, or put hands on their patients, I watched and made sure all protocols were being followed. I was not there to police, but to support and protect; while they cared for the patients, I cared for them. Still, I can tell you that if a biosafety officer is sitting in the laboratory watching a scientist work, the scientist will be much more likely to follow safety practices—an example of the effects of extrinsic motivation. This is not practical, however. Monetary rewards for those applying safer behaviors also are both impractical and not dependably effective. Again, it is important to point out that extrinsic motivation does work, but by definition it comes from outside the individual and can require substantial time and resources.

When motivation comes from outside an individual, one must ask what the individual will do when the motivating force is no longer present. Does the individual have any reason to behave other than the presence of the extrinsic motivator? If they do have a reason, do they know what it is, and how long will that reason exist with no reinforcement or support of the requested behavior? People become dependent on the extrinsic factor for behavior to occur.

People can be motivated by extrinsic forces. However, when it comes to safety, unless you have unlimited resources and time, extrinsic motivation is simply not practical or effective in motivating those working on the frontline with infectious diseases to practice safer behaviors.

SYSTEMIC MOTIVATION

It is a waste of time to ask if your organization has a culture of safety. Of course it does! It may not be the culture you want, but it is a culture. A culture is established by a common set of beliefs and behaviors within a specified group of individuals. This group establishes a set of social norms based on these common beliefs and behaviors. These social norms produce systemic motivators that, if not controlled for, can increase overall risk to health and safety.

A safety officer (extrinsic motivator) can educate about the consequences of eating in the laboratory. However, if every Friday is doughnut day in the laboratory, you would be considered an outsider if you did not join the team in having a doughnut (systemic). Rather than facing scrutiny and be ostracized, you eat a doughnut and are welcomed and accepted. A very powerful motivating force indeed!

Although extrinsic motivation can be useful, its practicality is limited with regard to motivating individuals to participate in safer behaviors. But systemic motivation can be very powerful. If leadership of an organization is involved in safety, the impact it can have on group behavior is substantial. I believe that leaders CARE, that is, they ensure compliance through accountability when providing resources and setting expectations. If leaders ensured that the workforce was attempting to live up to expectation (compliance), and they provided feedback for when the workers did or did not do so (accountability), safer behaviors would occur as a result of very strong systemic motivating factors. In short, you would create a culture centered around behavior first. With behavior come beliefs. If you have a group of people behaving and believing in similar fashions, you have a culture.

So, what am I saying here? Safety cultures are a result of sound leadership. By setting expectations and ensuring compliance and accountability among the workforce, leaders can systemically motivate individuals to behave. But what happens in the absence of leadership? This is the biggest problem we face today.

As stated many times in this book, the greatest risk in laboratory safety management is not the biological agent or the people working with the agent. The greatest risk is the culture of the organization. Most of the time, systemic motivation increases risk rather than decreases it. Those working on the frontline are motivated for different reasons than leaders and safety professionals are. Time, money, and efficiency outweigh reputation, fiscal responsibility, and even health and safety. I don't believe anyone wakes up and decides to hurt himself or herself at work. But I do believe that those working around a risk can, over time, become complacent and lose respect for that risk. That is a human risk factor and will occur no matter where you are in the world. My point is that an unchecked culture could be a risky one. Systemic motivation can work for or against you, and you should be aware of that. It may be more practical, but it requires leadership at the highest level to get the best results specific to safer behaviors.

INTRINSIC MOTIVATION

The real question is this: how do we prime a group of individuals to receive a certain amount of satisfaction when they practice safer behaviors? Intrinsic motivation comes from inside an individual. It is not a reward from the safety professional or recognition from the organization. Please understand that recognition and rewards do motivate people to behave. However, there is nothing stronger than a self-motivated individual. Someone who is intrinsically motivated can do just about anything—and justify the means for doing so!

A classic example of intrinsic motivation leading to risky behavior is cell phone usage. For a variety of reasons, cell phones provide certain levels of pleasure. This is so attractive that people will use cell phones while driving a car going 65 miles per hour. Some will bring cell phones into the laboratories or talk on them while treating sick patients. You know individuals are intrinsically motivated when you point out how risky the behavior is only to have them defend it, as if they can make the behavior seem logical and acceptable regardless of the risk it exposes them or others to. If we could get people to defend safer behaviors the way they defend the constant use of their cell phones, how much safer this world would be!

Taking intrinsic motivation and plugging it into a strategy called intrinsic safety is not an easy task. First, you must learn the skill of how to motivate individuals intrinsically. After mastering this skill, you must also establish practices throughout the organization that increase intrinsic motivation among people to practice safer behaviors. If you can put both of these together, you will increase the capacity and desire of staff to do safety for themselves. This is the overall goal of any intrinsic safety program.

Motivating Others Intrinsically

Being a parent is tough sometimes. For the sake of time and to end our children's frustration, we can be tempted to jump in and help them do something or save the day for them. In the end, however, what does that do for them? When something goes wrong or they get frustrated in the future, what have we taught them about how to deal with it? Rather than hunkering down and figuring the problem out by themselves, they might instead look to others for the solution.

To make this point, I tell the following story in some of the trainings we do. I hope it resonates with you as well. There once was a wonderful father who had a beautiful daughter. The father did everything he could to make sure the daughter had everything she needed. Unfortunately, this meant he incurred a tremendous amount of debt. The debt was held by a very ugly banker, ugly inside and out: mean and prideful and totally lacking in integrity. He cared more about what others thought of him than who he was as a person.

One day the banker decided it was time to collect from the father. The father could not pay the debt, and the banker threatened to throw him in prison. However, the banker offered to forgive the debt if the father would allow him to marry his daughter. When the father heard this, he elected to go to jail rather than allow that tragedy to happen. His daughter objected strongly to this, so the banker made a counteroffer: "What if I put a black rock and a white rock in a bag? If you pull out the black rock, you must marry me, and I will forgive the debt. If you pick the white rock, you do not have to marry me, and I will still forgive the debt. Either way, the debt will be forgiven and your father will not go to jail."

At this, the daughter accepted the offer. The father, in turn, objected, and she began to console him. As they hugged, she noticed the banker placing two black rocks in the bag. She saw now that the game was rigged, and she was going to be forced to pull a black rock out of the bag.

Her father had always told his daughter that for every problem there is a solution, but that those solutions do not come from outside. They arise from your own ability to remain calm, silence fears, and attack the problem logically. He assured her that she was capable of solving any problem, creative enough to develop a working solution, and competent enough to make it happen. Although at times she grew very frustrated at her father's belief in her, his support had made her a true problem solver, one who looked within herself rather than to others for solutions.

She thought for a moment. Then confidently she walked up to the banker, stuck her hand in the bag, and grabbed a rock. When she pulled her hand out she intentionally dropped the rock on the ground, and it was immediately lost among a bunch of other rocks. She pretended to be shocked and apologized, "I'm so sorry about that! I guess we will have to look in the bag and see which rock is left. That way we'll know which one I accidentally dropped."

She pulled the remaining rock out of the bag and showed everyone that it was black, from which everyone assumed that she must have originally grabbed the white rock. She was not forced to marry the banker, and he still had to forgive her father's debt. The banker, a proud man who would never admit to cheating, was forced to comply and walked away wondering how the daughter had gotten the best of him.

Here is my point in recounting this story. When a safety problem presents itself to a scientist, nurse, or doctor, will there always be someone there to help them? And do we want them to always look for help to achieve an acceptable level of safety? Teaching others to be safe for themselves is the best approach for the long haul. It may feel good to be needed, but when you are not around, and right behavior must continue to occur, it is best if people know how to be safe by themselves rather than wait for the safety officer to come and save the day.

How do we motivate others intrinsically? The first step is changing how we deal with people. Usually if someone comes to us with a problem, we listen, ask questions, and immediately begin attempting to solve the problem for them. We offer solutions that have worked for us in the past and discuss a variety of options from our perspective. But what does this accomplish? If they choose your solution and fail, you end up getting blamed and they learned nothing about how to solve future problems. If your solution worked, what happens the next time they need something and you are not there to assist? What can you do instead?

Step 1: Replace the word "help" or "fix" with the word "serve"

Let's start by shifting your perspective. Instead of using words such as "I am going to help you" or "I am going to fix this situation," begin thinking about how to serve them. *Helping* someone gives the impression that you think that they are not skilled enough to solve the issue on their own. *Fixing* something assumes that it is broken, and fixing someone means you are doing work for and to someone. *Serving* someone means you assume they are skilled, can bring viable solutions, and can be sensible in their overall approach.

Serving someone means that rather than doing something *for* or *to* someone, you will be doing something *with* someone so it can be done *by* that person. Plug in the word "safety," and it sounds like this: "I am not going to do safety for you or to you, I am going to do safety with you so it can be done by you." That is a statement of intrinsic motivation. Many of you have years of experience. You may have solutions that are waiting to see the light of day. You can solve problems and provide insights that bring value and could even save lives. I am not asking you to withhold your knowledge, I am simply asking you to use it differently. Rather than offering an immediate solution to a problem, why not approach it by asking the question, "What do you want to do?"

Let's say there has been a hazardous spill. We all know there are several ways a spill can be cleaned up safely. Before telling the staff responsible for cleanup what you know about cleaning spills, let them think about their environment, the resources they have, and those they work with and consider solutions they think would work for them in this situation. If the proper solution is their idea, they are more likely to follow it. But what if they run into a wall? Perhaps they say, "I think we should let the disinfectant sit on the spill for 45 minutes." What do you say then?

You shift from the intrinsic motivation to extrinsic motivation and say, "When we do a risk assessment, we consider the agent we are working with and the disinfectant we are using to inactivate the agent. Based on my experience, the disinfectant we are using will completely inactivate the spilled agent in less than 5 minutes." Then you switch back to using intrinsic motivation and ask, "Now that I have said this, what do you want to do?"

Serving means that you guide someone to a solution that works best for them, not for you. There is no doubt the process can be frustrating to those who are just learning about intrinsic motivation. Most of the time, people make a request, and the request is fulfilled for them. But we are not talking about a fast-food order; we are talking about safety. When it comes to safety, intrinsic motivation will encourage individuals to do safety for themselves rather than having it done for them. That means structuring programs that promote safer behaviors when no one else is looking. In other words, what you do when a safety official is watching doesn't matter nearly as much as what you do when the safety official is no longer around. What matters most is that you behave safely at all times regardless of whether someone is looking. We start to achieve this when we invite people into the conversation of safety and serve them rather than fixing the problem for them.

Step 2 discusses the beginning of the service process.

Step 2: Treat people as though they are skilled

Anyone can do safety. Please do not doubt that. In fact, our mere existence means that we are natural risk mitigators. We have the skills and abilities to identify hazards, assess the risks, and manage them appropriately. So, moving forward, if your goal is to motivate those you serve intrinsically, it's best to assume that those you serve are skilled in solving problems. I want you to believe they have the skills necessary to solve the challenge in front of them.

Please remember that you can shift from intrinsic to extrinsic motivation at any time, as in the example above. If individuals try and obviously lack the skills needed, you can demonstrate what is needed, mentor them, and allow them to practice the desired behavior (extrinsic motivation). However, you should always return to intrinsic motivation by stepping back, asking them to perform the behavior, and then asking them what they want to do now that they have the new skill.

Assuming that people have the skills to solve their own problems is a strategy that is not about them; it is about how you view others. If you choose to apply this concept, rather than providing the skills they need, you are asking them to explore the skills they already have while asking them to demonstrate the skills they will need to solve the problem. Your skills matter, but if you are not there when they experience a problem, their skills will matter more when it comes to safety.

Step 3: Treat people as though they have the solution

Someone comes to you with a safety problem: he is having problems with staff using cell phones at inappropriate times. When you apply step 1, you put yourself in a position of serving by facilitating the process of allowing him to solve this safety problem. You choose not to do this for him or offer your solution to his problem. Instead you say to him, "I believe you have what you need to solve this issue (step 2). I also believe that you can come up with solutions. What are you thinking about doing?"

Trust me—this is the hardest part of the process. It's natural to want to jump right in and offer a solution, solve the problem, and make their life easier. It's the easier thing to do, but it does nothing to help the person with the problem. Your solution works for you, not them. The harder route is to guide them to a solution that works best for them. Keep assuming that they already have a solution within them, but the two of you may have to work together to discover it.

I usually try not to make assumptions about others, but you need to do it to intrinsically motivate others. It is this assumption that brings out the best in those you serve. As long as you are serving with them and by them throughout the process, you have a chance to either confirm or reject your assumption. Either way, the process of intrinsic motivation is about controlling your urge to solve their problems by using your experience and solutions. Instead, you should use your experience and solutions to guide others to develop their own.

Step 4: Treat people as though they are sensible

As I mentioned earlier, humans do not waste their energy doing things that offer little or no benefit. Assuming those you serve are sensible, meaning that you know they will apply wisdom in a way that is beneficial, is pretty safe, to a certain degree. Unfortunately, there is some bias here that has to be controlled. Some scientists and health care providers may be more interested in doing science or treating a patient than they are in practicing safer behaviors. In these cases, it's not because they don't want to be safe, it's just that the safer behavior has not demonstrated a benefit, so they will not choose to do it.

Once an individual develops a solution, it is your job to assume that he or she will implement the solution in a sensible fashion. Again, you must serve by doing safety with people as they begin implementing the solution they have developed. Sometimes people who implement new programs need assistance because they run into challenges that may be over their heads to solve. In that case, serve them by providing perspectives (extrinsic motivation) but always return to intrinsic motivation by the question, "What do you want to do?"

Step 5: Continue asking the question

Believe it or not, when you get to this point, you have already done the hard work. The rest is easy and requires the skill of listening. When someone comes to you with a problem or a need, ask him, "What do you want to do about this?" If you ask this question properly it can produce new thinking. New thinking leads to new behaviors. Instead of doing the thinking for them, you are facilitating the process of letting them think for themselves. They will more willingly implement their own solution than something prescribed by another.

If people respond to the question by saying, "I don't know. Can you tell me what I need to do?," I suggest you respond with something like, "I could tell you what has worked for me and others, but this is about you. Take some time and think about this. What are you thinking about doing specific to this situation?" This will take some practice and you may be skeptical. It can be frustrating for everyone. It's much easier to direct people if you think you know more than they do and what is best for them. However, using this technique evokes better and more innovative solutions to safety problems.

It can be concerning and challenging, however, when new thinking begins to occur. You might hear ideas and ways of thinking that can scare and even shock you. It does not matter if what comes to their mind is ridiculous, impractical, or even wrong. What matters is that they begin using their minds to implement their skills, solutions, and sensibility to address the challenges they are encountering. When they start thinking, we can start serving. Let them guide the process, and be patient with the pace at which they choose to go. Remember, you are not there to save the day. You are also not there to fix something that is not really broken. You are there to serve by being with them throughout the process. You will see that they will get to where they need to go, although they might use a different path than the one you took. It helps to think about what you want to do specific to serving others in safety right now.

When you practice intrinsic motivation, you start by changing your perspective. Your goal is not to see yourself as better, more experienced, or a fixer of problems. Instead, your goal is to serve people through a process that allows them to demonstrate their skills, solutions, and sensibility. This comes by asking questions, listening, offering an observation, asking more questions, listening again, holding back things you want to say, listening more, encouraging, affirming, waiting, offering another observation, listening still more, reinforcing, and then…discovering! You will witness staff begin to talk themselves through problems and then solve them on their own. This demonstrates their increased overall confidence and skills to behave safely when faced with a challenge.

Signs That You Are Practicing Intrinsic Motivation

When I first started serving, it was very challenging. I am an extremely extroverted individual. I like to talk, I love to help, and I don't like watching anyone struggle when a solution could be easily offered. But realistically, I cannot be a safety knight in shining armor who comes to rescue everyone. Doesn't this mentality make their issues more about me than about them? Who gets the recognition for success? Who gets the recognition for failure? True, there are benefits to extrinsic motivation (solving the problem for them): it is quicker and easier. But extrinsic safety builds dependency instead of their capability to be safer without you there. Remember that as you experience how it feels when you start intrinsically motivating others in safety.

Lots of silence. When you first start serving others in this capacity, you will have to become very comfortable with silence. There will be lots of silence. Try not to view this as something negative. In fact, silence is a great indicator that someone is thinking. Many of us view silence as awkward and rush to fill it with something. But people can't think if you are talking. More importantly, when you are filling the silence, what message does that send? When you have challenged someone to provide a solution, and there is silence, you might panic, feeling as though you have asked too much of them and now they seem to be struggling. Your urge to help or fix grows even stronger now, and maybe you begin shooting ideas at them, rapid fire. Try to resist.

Struggle is growth. You don't gain muscle watching others lift weights at the gym. You also don't gain muscle listening to how others gained muscle lifting at the gym. When you lift weights, you struggle; when you struggle, you are forced to go within and learn; when you learn, you change. Because that change comes from within, it motivates through the good feelings people get from doing it for themselves.

Lots of "I don't know"s. The use of technology has minimized the use of three words that motivate learning at the highest levels.

I

(me, my experience, my knowledge and beliefs)

DON'T

(an absence, void, lack of)

KNOW

(belief there is a lack of knowledge and experience)

When you ask someone a question and they say I don't know, letting them think begins new thinking, and new thinking leads to new behaving. When I conduct a safety training program, I always start with a pretest that is designed for individuals to fail. Unless they have previously attended one of my training programs, they cannot possibly pass this test. It's a method for priming a person for a learning experience. The words "I don't know" are a personal declaration that learning is needed. When someone comes to understand this, it can be the single greatest moment of intrinsic motivation. What people do in this moment will determine how they grow. Will they become more dependent on others or turn to themselves?

Hearing the words "I don't know" signals that intrinsic motivation is starting. It is perfectly fine if people don't have a solution immediately for any safety issue or problem. Encourage them to take time to think about the issue. Let them think out loud with you or return later to discuss the issue more. The worst thing you can do when someone tells you they don't know something is to tell them what *you* know.

Intrinsic motivation is becoming harder and harder because of technology. Google allows us to search for solutions to common problems. Again, I acknowledge that extrinsic motivation works. It can solve problems quickly and efficiently. However, dependency on something other than yourself to solve your problems should be alarming when it comes to safety. Safety is a collective effort that begins at the individual level.

Safety is the outcome of plans plus individual behavior. When someone doesn't know something, you need to believe that the person is skilled, can develop a solution, and is sensible enough to do so without the Internet. Give them that challenge; don't run away from it. Silence and "I don't know" are indications that you are serving, because when someone is invited to think, they must be silent. "I don't know" is what usually fills that awkward moment.

Lots of urges to jump in. I still struggle with this. I can get people to start the thinking process. I can become silent and listen for them to say, "I don't know." But then I want to jump in and give them a solution that seems right there in front of them. There is a famous quote from Gandhi that summarizes the intrinsic process completely, which I have adapted for all genders: Give a person a fish and you feed that person for a day. Teach a person how to fish and you feed that person for the rest of his or her life.

Reminding myself that a controlled struggle promotes growth usually allows me to step back and watch the process unfold. If you come to the rescue, when people are challenged in the future, they may look to you to make another save. The urge to jump in, offer suggestions, and solve problems should be resisted until such intervention is absolutely needed. Only when someone is about ready to give up and quit should we offer a thought. Having done that, we should ask them again what they want to do. Keep assuming that everyone you are intrinsically motivating has the skills, solutions, and sensibility to mitigate a risk or offer a safer alternative to the challenge they are facing.

Using this strategy can be a little risky. In the beginning, you may frustrate people. If they come to you, they generally want answers right then and there. They may ask you to solve the problem for them. It may take some people more time than others to think for themselves. However, I have witnessed the value of this strategy. It increases safety tenfold when individuals solve problems the way they would, not the way you would. For example, when people decide exactly how they want to behave, they connect the situation with a preferred behavior, which makes it more likely for the behavior to occur. However, when you decide what they should do, that connection does not necessarily occur. Therefore, you should facilitate the process (the connection), never leave them alone, and serve them until the connection is made, safer behavior is identified, and the process concludes.

Intrinsic Safety Program Examples

Now that we have explored how we can influence a person's individual behavior and motivational drive, what does intrinsic safety look like at the organizational level? Safety officers have a chance to motivate intrinsically those they serve. This can have an impact at a personal level, but what can be done to build a culture that promotes intrinsic motivation within safety practices?

Here are examples that can be immediately implemented within organizations to increase the ability of the workforce to practice safer behaviors. There might be resistance at first, but in the end there will be increases in ability. This minimizes the dependency on the safety office to identify and manage risks for others. The following things can happen when there are direct requests for a partnership between leadership, workforce, and safety officials.

Safety official reviewing standard operating procedures, not writing them. If a plan specific to safety has to be written, it is important to stress that it should be written by the team practicing the behaviors, then reviewed by the safety professional. The plan may be the worst thing that you have ever seen as a safety official. It probably will be. But let's review a little psychology. First, if someone drafts a plan on paper, they are more likely to change than if they just talk about or agree to something. Second, bringing a group of people together to discuss and develop a plan of action to improve safety is the way to develop a culture. When a group of people within a specific unit address general risks by establishing common rules and observable rituals, a culture is storming or forming! Finally, asking those outside safety to come up with ideas on how they can do what they want to do safely is real training for those working with risks.

Plans should be reviewed, but not written, by safety professionals. When a safety professional identifies a concern, their response should be along the following lines: "I like this plan overall, and I have some concerns. For example, on step 12 you state…"; "Have you considered…?"; and "Now that I have shared that, what do you think you want to do?" This process teaches the workforce to think more like a safety official and puts them in control of establishing their own standard operating procedures (SOPs), or as I like to say, their own house rules. Writing plans for people fails to increase the ability for safer behaviors in those you serve. Request a plan, review the plan, and shape the plan—together.

Safety official serving rather than policing. I do not believe that safety and regulatory responsibilities should be mixed. I am a big believer in compliance and accountability. However, when a safety official becomes legally bound to report minor incidents and accidents, their ability to serve staff members becomes inhibited. You cannot serve both the workforce and the regulatory body. It may seem that by doing one, you can do the other, but regulations can bind you to specific actions that do not allow control to remain in the hands of those you are supposed to be serving.

Intrinsic safety programs ensure that safety officials serve the workforce with strategies that include reinforcement, appreciative inquiry, and positive accountability. In other words, although a safety official has the right to focus on what is not going right, instead the focus should be on what the workforce is doing right. People do not go to work to try and hurt themselves or others. Safety is inherent in sane people. You may witness complacency, apathy, or fatigue, but having the intention of hurting oneself or one's coworkers is extremely uncommon. My point is, why not focus on what is working rather than what isn't?

Intrinsic motivation focuses on motivating an individual to behave. Punishment aims to stop behavior, which is typically the outcome of regulatory requirements. Management systems and regulations, such as the Occupational Safety and Health Administration and the Federal Select Agent Program, have measurably changed safety levels within organizations. But have they increased ability or transparency? What safety metrics have they used, and how have they demonstrated safer practices?

I believe you can get to safer behaviors with serving rather than policing. Can we intrinsically motivate someone to behave by policing? I don't believe that is possible, because the goal of policing is to stop behavior rather than start it. When you are serving others, the goal is to start new behaviors with new thinking. When you are bound by law to serve in a specific manner, you are serving the regulatory body, not the workforce.

Safety audits facilitated by the workforce rather than the safety officer. Let's imagine you have a choice. Would you prefer that (i) I send someone to your house to inspect it for safety issues, (ii) I give you a checklist to inspect your home for safety issues, or (iii) we both inspect your house together after completing the same checklist independent of one another? Which option do you think would increase the capacity of the workforce to realize when a risk surfaces in their work environment?

If a safety audit is done independent of the workforce, does that teach the specific skills needed to identify and manage risks? Give this a try. Create a safety checklist. Send it out randomly to 30% of laboratories or health care facilities. Request that department heads facilitate a safety inspection using the provided checklist. Gather the results, and within a week have the safety office visit the departments and inspect them using the same checklist. Compare the survey results with the inspection results by reviewing the checklists, with an emphasis on the differences of perceptions. This process aims to begin bringing one perception, that of the workforce, in line with the perceptions of the safety officer.

It does us no good if those working around risks cannot see risks unless the safety professional is there. It is important that the safety professional teach those who work in laboratories and health care facilities what constitutes a risk from the perspective of a safety professional. Additionally, the safety professional will understand the perspectives of those they serve. Either way, this process should be one of learning, positive and reinforcing. It should not be a scolding session or one that scrutinizes or embarrasses the workforce.

The more the workforce audits itself and safety officials review those audit results by comparing results, the more likely perspectives of safety and science will be blended. When these two are blended and become one, you have safer science done with and by scientists, rather than for or to them.

The workforce, not the safety officer, presenting to the safety committee.
There is a common trend to ask safety professionals to present protocols
meant for the workforce to the safety committee. What does this do? It removes the workforce from the safety responsibility. Rather than excusing
the workforce from this review process, why not have the safety professional review and prepare the workforce for the meeting? This does two
things. First, it teaches the workforce to involve safety in the development
of and review of protocols. Second, it solidifies a sound relationship between the workforce and safety official.

Ask the workforce to present safety protocols to the safety committee.
This allows the committee to ask safety questions of the workforce directly.
This will also ensure higher levels of compliance and accountability. Why?
When people publicly state that they are going to do something, as op-
posed to having someone do it for them, they are more likely to do the
behavior. In this way, if they don't behave in the desired way, they can be
held accountable by the entire committee as well as challenged by peers
about why what they agreed to was not being done.

Signed behavioral expectation contracts. Last and certainly not least, asking the workforce to sign a behavioral expectation contract annually can
intrinsically motivate them to behave correctly. I have mentioned examples of behavioral expectations throughout this book. They are not like a
manual of SOPs. Instead, they are simple, clear, and easy to measure
through observation. Because we are asking people to behave in specific
ways, we can observe them to see if expectations are met. If the behaviors
are present, they are meeting the expectation. If the behaviors are not present, they are falling short of expectations.

Safety should have much more than guidelines, policies, and SOPs.
The workforce should have a clear set of expectations that decrease risk
through a set of clearly identified behaviors. Putting these behaviors in a
contract, expecting workers to comply with the behaviors, and holding
them accountable for doing so or failing are imperative for ensuring that
you have created an intrinsic safety program.

Intrinsic motivation changed how I did things. I used to do my best to
solve everyone's problems. In doing so, I learned that I was not serving
them but increasing their dependency on me. That is what intrinsic safety is
all about. Rather than placing the sole burden of safety on the safety professional, the workforce is included to do safety with the safety professional,
which increases their ability to practice safer behaviors. If we do safety for
them or to them, they will defer to the safety professional to be safe.

The workforce must learn to regulate itself for two very important reasons. First, it is the most practical strategy, saving money, time, and lives.
Second, each person in the workforce is responsible for safety. Because
safety professionals cannot always be there, the workforce must learn to be
safe for themselves.

BEHAVIORAL EXPECTATION CONTRACT

LABORATORY ACQUIRED ILLNESSES (LAIs) POSE A SERIOUS THREAT TO LABORATORY STAFF, SCIENTIFIC REPUTATION, RESEARCH FUNDING AND THE ORGANIZATION AS A WHOLE. INDIVIDUALS WORKING IN BIOLOGICAL LABORATORIES (BSL1, BSL2, BSL3, AND BSL4)) ARE EXPECTED TO PERFORM WITH THE HIGHEST LEVELS OF INTEGRITY AND COMMITMENT TO ORGANIZATIONAL SAFETY AND SECURITY POLICIES AND PROCEDURES.

BASED ON SEVERAL HISTORICAL CASES WHERE LABORATORY STAFF BECAME ILL AS A RESULT OF THEIR WORK IN THE LABORATORY ENVIRONMENT, THE FOLLOWING BEHAVIORAL EXPECTATIONS MUST BE AGREED TO.

1. I WILL FOLLOW ALL STANDARD OPERATING PROCEDURES TO THE BEST OF MY ABILITIES.

2. I WILL ENSURE OTHERS FOLLOW ALL STANDARD OPERATING PROCEDURES TO THE BEST OF THEIR ABILITIES.

3. I WILL REPORT ANY LABORATORY NEAR MISSES, INCIDENTS, OR ACCIDENTS.

4. I WILL REPORT CLINICAL SYMPTOMS WHICH MATCH THE CLINICAL PRESENTATION OF ANY PATHOGENS WHICH ARE BEING WORKED ON IN THE LABORATORY.

5. I WILL REPORT ANY NEW MEDICAL CONDITIONS WHICH MAY AFFECT MY SAFETY OR THE SAFETY OF OTHERS WHILE WORKING IN THE LABORATORY (INCLUDING BUT NOT LIMITED TO DIABETES, HEART DISEASE, ACUTE ASTHMATIC CONDITIONS, MEDICATIONS CAUSING SEIZURES OR COMPROMISED IMMUNE SYSTEMS, PREGNANCY).

BY SIGNING THIS, YOU UNDERSTAND AND AGREE TO ADHERE TO ALL LABORATORY STAFF BEHAVIORAL EXPECTATIONS.

NAME: _____

SIGNATURE: _____

DATE: _____

Figure 9.1 Behavioral expectation contract

BIOSAFETY *in the First Person*
The Bruises of Biosafety by Sarah Ziegler

I first met Sarah Ziegler when she was an NIH NBBTP (National Biosafety and Biocontainment Training Program) fellow. Energetic, analytical, and strong, she made an excellent first impression on me. I made a mistake with Sarah. At one of the locations where Sarah was working, there was a great deal of anger. In a public forum and with Sarah present, I asked individuals to openly share this anger. I believe this hurt Sarah and for that I remain apologetic. However, using this example, I want to let each of you know something very important. Your value is not determined by what others think about you. Nobody in that room with their words of anger determined Sarah's value. Regardless of being thrown down, stepped on, and mistreated, Sarah remains an extremely valuable asset to the profession of biosafety. No words can ever take that away from her.

Biosafety for me is the most natural career I can have. This does not mean that I don't have challenging days or that I haven't made mistakes, but for me, every aspect of this profession just makes sense. I see the biosafety anytime I walk into a space. That's hard to explain sometimes. It's kind of like driving on the right side of the road, something we don't even have to think about, we just do it. Biosafety is like that for me: I don't really think about it any more, I just do it.

For as long as I remember, I have been a scientist. I used to collect pond water in the backyard to look for living things with a microscope. I even started doing molecular biology at age 14, with things like liquid chromatography and get electrophoresis. So, yes, I was that nerdy kid. When I got to high school, I was determined to save lives by trying to cure diseases. I was going to cure HIV and cancer, just to start. I followed a fairly typical path through academia, but then the movie "Outbreak" came out. It was the first time I had ever seen anything like the BSL4 suits and lab environments in the movie. I knew that it was not real, but I became enchanted with the idea of working in high containment. I applied to University of Texas Medical Branch (UTMB) when a friend of mine called me from campus saying, "You are not going to believe this, but there is a level 4 here. You have to come." I worked in high-containment labs throughout my graduate degree and postdoc while at UTMB.

By the time I finished graduate school, I had become slightly frustrated with science because I realized that I might not actually be able to cure any diseases or save lives. A position was open in the biosafety training program at UTMB, and I was interested. I had always worked well with all my biosafety team. I think it took about two months for me to realize that biosafety was perfect for me. I loved the people, the problems, the science—really, everything about biosafety.

I am extremely lucky to have been accepted into the National Biocontainment Training Program (NBBTP). NBBTP is like getting thrown into the deep end of the pool without any lessons. It was scary and fast paced, but what I didn't realize at the time was that none of my mentors were going to let me drown. The opportunities in the program were endless, and you just had to dream and ask for it to come true. I learned from and trained with the best in the field. I had the opportunity to travel the country and learn many ways that biosafety can be successful. I credit all my past and future success to these great people, who have generously given me guidance throughout my career. I would never be as successful as I am today without the NBBTP.

How do you learn to be a biosafety officer? I think the best way to learn is by doing and shadowing. There are many scenarios that you need to be able to understand, as well as prepare for, that you cannot learn in a book or on the computer. Others may think that all we do is inspections and training, because that is when they see us the most, but that is less than 10% of my job and the easiest to complete.

As a biosafety officer, you somehow have to find your crystal ball. I imagine the worst-case scenario for everything now, even in my personal life. If I see an armadillo crossing the road, I think about children contracting leprosy. The bats flying around the yard at night may be neat, but I think they carry rabies, or maybe even Ebola. In my labs, worst-case scenarios involve death, maybe one person, maybe thousands. A coworker of mine was doing a business risk assessment for her division, and she wanted my input. It was a typical risk assessment, with a list of potential hazards with the likelihood and consequences scored numerically. She had listed some things as having consequences that were worst-case scenarios, 5 out of 5. I asked her what were these horrible consequences, and she replied losing money, losing clients, or maybe harming the company reputation. I looked at her shocked, and replied, "But nobody dies, so how is that your worst-case scenario?" She laughed, and said that this wasn't a biosafety risk assessment, it was for the business. Biosafety has given me an interesting perspective on life. Dryers breaking, credit card lost, a dead battery—these are no problem. I worry about outbreaks, hurricanes, mass casualty events, and how to get an unconscious person out of a high-containment lab and transported to a hospital without exposing anyone or the environment to deadly agents.

Although biosafety may be natural for me, the days can still be difficult. I think this is because I love it so much that or the thought of failure is unbearable. In biosafety, you are the ultimate middle man, and you are never going to make everyone happy. Debbie Wilson once told me that if everyone liked me, then I wasn't doing my job correctly. We are sometimes the spokesperson for our organization, defending it to regulators and championing our researchers' science. But sometimes, we are the inspector of our labs, pointing out the flaws and pushing for drastic changes. We wear many hats during our day-to-day dealings. We may be in meetings with the CEO asking for money or explaining problems, and then we may train facility staff about proper cleanup of human blood.

I have now spent most my career fixing things and learning from mistakes. It fits my personality, to fix things and be challenged like this. I have found that fixing the biosafety challenges is the easiest part. Writing a standard operating procedure (SOP), making a new form, and optimizing a space are the easy challenges. It's the people and the culture that are the hard parts. They can be frustrating and can make you lose your focus. At my core, I do my job to save lives and make sure my researchers are safe. It's not about rules and regulations. I want people to leave work every day in the same health as when they walked in. It is hard to hear from the people that I am working every day to protect that they don't see value in what I do. I have debated endlessly with regulators to allow a primary investigator (PI) to do his research and then had that same PI accuse me of trying to ruin his work. It is almost heartbreaking when this happens. You begin to question your purpose. That is the hardest part of doing biosafety, having to fight for the same people who are sometimes actively fighting you. It's thankless. You don't usually get thanks or even an apology. The lab workers rarely see the dangers that you are trying to protect them against. And unfortunately, if something catastrophic happens, you are to blame, because you didn't do enough or anticipate that problem.

It's the people and the culture that drive good biosafety. It is not the written SOP or the form you create for checking things off; it is how the people interact and respect safety that dictates how successful your program will become. If you have good people working together, SOPs become secondary because everyone understands what is the best practice. You don't need forms to double-check processes or reviews, because there is compliance with SOPs. But the people aspect is the hardest to build and change. Unfortunately, one bad apple can poison the whole barrel and upset any progress you made. And don't forget those people who have been doing it a certain way for years, because it will feel like years to get them to change anything. Culture change is hard and mostly unpredictable, but when you can manage to do it, it may be the most rewarding accomplishment.

How do you measure success in biosafety? Unfortunately, it is much easier to measure failure than it is to measure the success. A failed inspection or a horrific injury can show how something went wrong, but how can you tell if everything is going well? There is no perfect score card for biosafety. Honestly, when things are going right, you are the most forgettable person in the organization. Fewer incidents means fewer calls, and if you train the staff right, they should be able to manage their safety basically by themselves. So, your reward for success is silence, but during an incident you become the most important person in the room or organization. It is no mystery why people get nervous when the biosafety officer makes an unannounced visit;

we rarely bring good news. At one point in my career, the assistant to my su-pervisor stopped saying hello and just greeted me with, "What now, and how bad is it?"

Don't get me wrong—biosafety is fun, too. One of my first memorable experiences as a new associate biosafety officer was when I got the call to help out a lab. They had found a monkey head in a refrigerator and needed it transported to pathology all the way across campus. I still remember carrying that cooler those few blocks. I've had amazing experiences throughout my career. I helped train a medical team to respond to the Ebola outbreak in Africa. I have trained with a bomb squad, the FBI, firefighters, police, and emergency medical technicians. For some reason, I end up on roofs a lot to look at ex-haust fans, and I have learned not to wear heels or skirts without a change of clothes in my office. I am trained to wear full hazmat gear, and when alarms sound, I am the person who runs into the building, not out of it. Now if you read this and think that none of this sounds like fun, then biosafety may not be for you. There are boring parts of the job, too. Paperwork never ends. Counting tens of thousands of tubes can take days. I've watched lab workers pipette for hours and had trouble staying awake, and my calendar is usually full of back-to-back meetings.

Another great aspect about biosafety is the community. When you are having a bad day, or a bad month, the small community of biosafety profes-sionals is always just a phone call away. Although biosafety may seem diverse, it is almost exactly the same in many places. Rough inspections, difficult sci-entists, facility failures—we have all had to deal with them at some time or another. Biosafety is a close-knit group of professionals who are constantly networking. We are spread across the country, but e-mails, phone calls, and the annual meeting are how we stay in touch with each other. I recently re-ceived my Registered Biosafety Professional certification, and I am number 478, meaning that fewer than 500 people in the world have the same accred-itation in biosafety that I do. That is fewer people than my graduating class in high school. It amazes and humbles me that I have been allowed into such a selective and elite group of professionals.

I am in the early stages of my career, and I often look forward to how I will make my mark on the profession. I hope that I will look back someday and see that I have had a positive effect on the world of biosafety. Maybe I will help write a future version of the BMBL or be someone who develops the BSL5 lab. What I think will be the most rewarding is to know that I helped mentor a new generation of biosafety officers. I know that my men-tors and colleagues have helped me, and continue to help me, grow and learn. I want to pass that experience on. I hope that in the future, others will be able to look at me and say that I helped them become better biosafety professionals.

chapter 10

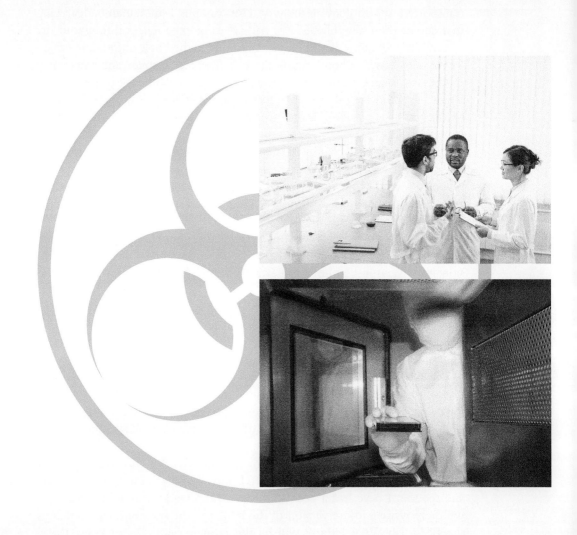

Building a One-Safe Culture

A *culture* is a group of individuals, belonging to a specified unit, who share a common set of beliefs and behaviors. A *climate* describes a group of individuals, belonging to a specified unit, who share a common set of beliefs and behaviors, but believe and act in their own best interests as individuals.

Visualize a huge iceberg. What you see above the water is only about a third of the actual iceberg. The part you can see can be compared to climate, and what is beneath the surface, unseen, represents the culture. The greater the distance between the top of climate and bottom of culture, the greater the risk an organization faces.

The difference between culture and climate is a very important distinction. Talking safety and doing safety must be aligned; it is not good enough to just say it, you must do it! How do we develop a safety culture, as defined above? First, we must identify the existing cultures. Even if your organization doesn't have a formal safety policy, what you have is still an existing culture of safety. It may not be healthy or liked and may even be hard to define, but it exists. You also have cultures of the workforce and leadership. Each of these cultures has different perspectives, and they may not trust one another. If they cannot trust one another, they will not work together. This poses a significant challenge to the development of a strong culture of safety. Rather than working together, the individual perspectives work in silos, which can make safety impractical and harder to do.

What, then, is a One-Safe culture, as I define it? Inspired by the concept of One Health—the blending of several different professions (including health care, environmental health, public health, animal health, and plant health) for the sake of identifying, preventing, responding to, and controlling infectious disease outbreaks—I developed the concept of a One-Safe culture (see chapter 28 in Wooley DP, Byers KB, *Biological Safety*, 5th ed, ASM Press, Washington, DC, 2017). This concept blends different positions (leadership, workforce, and safety) within an organization who subscribe to the fact that risk affects everyone in the organization equally and, thus, that safety is a common concern. As a result, the many are united into one, making safety a team effort where nobody is above anybody else. The group produces organizational rules to deal with the recognized common risk, which lead to predictable rituals not among individuals but as a One-Safe organization. What I promote is establishing a One-Safe culture, in which existing cultures and different perspectives are blended, expectations are set, and people are held accountable for meeting the established expectations. This is how trust is developed, because everyone knows what is expected of everyone.

The expectations for each perspective should be clarified, to increase trust within the culture. These expectations should be public knowledge so that individuals with different perspectives are able to regulate each other. For example, if leadership is not living up to its expectation, the workforce and safety should be able to challenge it, even though it is difficult to challenge authority; it is hard for players to confront a coach who is not doing what he is supposed to do.

If your organization says that annual review of standard operating procedures (SOPs) is being performed, but that is not actually occurring, this is a safety *climate*. When institutions can simply change the date and sign the same SOP that has been on the books for five years or more, this is a statement that there have been no behavioral improvements to the SOP, no changes, no updates. Can we believe that people are behaving the same way after five years? Another example of climate versus culture is the use of cell phones in the work environment. Almost everyone knows it is not a good idea. In fact, many organizations have established policies to keep electronics outside of the workplace, but, alas, they are still being used. What is worse is that the behavior is being hidden. It comes down to this: the more we know about what people are really doing, the safer those people are.

The most obvious hazard in laboratory and health care settings is not actually the greatest risk. The biological agent is the most obvious hazard, and those working with the biological agent are another obvious hazard. However, the greatest threat can be the culture of an organization. Although I am hopeful that more and more organizations understand this, I am beginning to see the development of climates rather than cultures. Organizations must commit to creating and maintaining cultures of safety. When the gap between what we say we do and what we really do is minimized, we live safer, healthier, and happier lives. This is an absolute truth and one that can easily be applied to safety.

I keep stressing the definition of culture because it is very tough to get individuals with different lines of work (e.g., science, safety, leadership, and clinical care) and with different levels of education and experience to share a common set of beliefs and behaviors. However, it can be done.

THE THREE Rs OF A ONE-SAFE CULTURE

I recommend a process that I have witnessed to produce a One-Safe culture in families, teams, and even organizations. Safety offers the opportunity to unite existing cultures and blend them to improve the overall organization.

Common Risks

Whether you are a scientist, doctor, nurse, leader, or safety professional, you share a common risk with others in your organization: if your organization has a safety accident that causes a serious injury or loss of life, it is going to have a profound impact. It could directly affect the health of people in and around the organization. The organization might have to recover from problems with reputation, closure of laboratories, suspension of certifications, or other challenges. The workforce, leadership, and safety professionals share a common risk to the continuity of the organization.

The next time you have members of leadership, safety, and the workforce together in a room, ask them three questions:

1. What do you think is going to be the next safety incident?
2. How would this safety incident impact your current duties?
3. Do you think you could prevent this safety incident by yourself?

The answers to these questions will produce a summary inclusive of all three perspectives, which ideally can be blended into one. When leadership, safety, and the workforce truly understand that they are connected because they have common risks, they identify as a unit, and the blending begins.

Common Rules

As mentioned in earlier chapters, if choice is removed, behavior can create beliefs even among people who have different perspectives, education levels, and experience levels. SOPs are a primary control of biosafety because they remove choice and provide rules for behavior around biological risks. It does not matter if you are a member of leadership, safety, or the workforce, when behaving around biological risks, you must follow the SOPs regardless of your perspective, experience, or education.

I have witnessed that compliance with and accountability to a specific set of rules lead to the establishment of common beliefs and thereby produce a safety *culture*. However, if rules are provided, but noncompliance and accountability are not addressed, the development of a safety *climate* occurs rather than a safety culture. In a safety climate, people will be able to tell you what the rules are even though they know they don't follow them. To achieve a culture of safety, we must establish common beliefs by ensuring that everyone in the specified unit follows the common rules.

The rules (SOPs) are important, but even more important is ensuring that the rules are being followed. The common risk brings people together as a unit, and compliance with rules ensures common beliefs.

Common Rituals

It may sound crazy, but I have often dreamed of advising a large organization that has leadership support that would allow me to establish such a strong safety ritual that people could tell, based on observing a person's behavior alone, where that person worked. I am honored to hear inspectors state that they have seen individuals practice the "beaking" method of glove removal or that they can tell in other ways when I have visited an institution. That means I have established common rituals in safety within that specified unit to produce good results.

Rituals are behaviors. Behavior can be observed and is the direct outcome of common rules. After all, rules typically don't tell you what to think; they tell you what to do. If you are asked to follow rules multiple times a day, a ritual begins to develop. However, is it the right ritual? Does the ritual match the rule? If it does, then you have a safety culture. If it does not, you have a safety climate—but if, and only if, the people can describe the rules. If they cannot even describe the rules, both safety culture and climate are missing. This is another reason to blend leadership, safety, and the workforce. Only when we bring different perspectives together to identify common risks, develop common rules, and produce common rituals do we form a One-Safe culture.

ESTABLISHING EXPECTATIONS OF LEADERSHIP, WORKFORCE, AND SAFETY

Each perspective has an important role within the specified unit that works around biological risks. Leadership is responsible for ensuring that the reputation of the organization remains intact. That reputation ensures the fiscal health of the organization and provides employment for the workforce. The workforce is interested in getting the job done efficiently and effectively, as their reputation and professional future depend on this. Finally, safety is responsible for ensuring that everyone within the unit remains safe and in regulatory compliance. As you can see, the pressures placed on each perspective mean that much is riding on a collective effort. Trust among the different perspectives is very important and begins with basic expectations.

When I have asked those I have served, "What are the expectations of you here, within the workplace?," I have received some surprising answers. Very few people have had a good answer for me. Then I ask, "How do you know if you are living up to expectation?" A shrug of the shoulder means we move along to the next topic.

The following is a starting point for the expectations for leadership, workforce, and safety professionals. This list is not a "be-all end-all" list, but if these expectations were implemented today, it would have immediate and profound impacts on safety within an organization. When I say profound, I mean noticeable and meaningful at all levels.

Leadership Expectations

Recently I have been begging anyone who is contributing to the 6th edition of the BMBL to include leadership as a primary control of biosafety. I am convinced that in order for safety to occur, the workforce and leadership must be completely engaged. Unfortunately, this control is usually termed "administrative," which dismisses leadership involvement in safety altogether. Just change one word, "administrative" to "leadership," and there is an immediate impact. When someone does not receive training, it is a leadership failure. When SOPs are not updated, it is a leadership failure. When medical or incident surveillance is not occurring, it is a leadership failure. Simply changing "administrative controls" to "leadership controls" sets expectations for leaders to engage in these areas.

But what do we expect of our leaders, specifically the ones who are leading the workforce and safety professionals? Here is a list describing leadership expectations for those serving around biological risks. As they are valid for almost any risk, they apply to other professions as well.

Leaders prepare the workforce. Leadership must prepare the workforce to do safety. Without adequate preparedness, both the workforce and safety officials cannot develop or sustain a culture of safety. Preparing the workforce comes down to three main tasks: (i) setting expectations, (ii) providing resources, and (iii) mandating training.

Leaders must set expectations and determine the general direction in which they want a group of individuals within a culture to go. The establishment of expectations is the start of holding individuals within a culture accountable. If workforce and safety professionals have a set of expectations, leaders can determine whether those expectations are being met. This accountability is crucial for ensuring that common beliefs and behaviors occur among those with different perspectives. Leadership has the right to expect the workforce to exhibit specific behaviors, but the workforce has the right to expect leadership to prepare, protect, and promote them by holding them accountable to organizational expectations.

Once expectations are set, leadership must make sure both the workforce and safety professionals have what they need to meet these expectations. Leadership must make sure the workforce has effective plans that minimize risk around infectious diseases. Leadership must also ensure that the workforce has access to proper and adequate equipment. This includes well-engineered laboratories, health care facilities, and personal protective equipment. Additionally, leadership must oversee surveillance and vaccination programs aimed at improving the overall safety program within the organization. It is not enough to simply provide expectations. Leadership must also provide the resources needed to work safely.

Finally, leadership must mandate training. As a parent of three, I can assure you that if my children were given a choice between playing video games and going to school, their choice would be playing video games (at least at this point in their lives). Science and care of patients are comparable to video games for scientists, nurses, and doctors. Why on earth would they attend safety training if they could be doing science and saving lives?

Leadership understands the numerous human risk factors that exist. These human risk factors require training as a control. Leadership must mandate frequent safety training for the workforce—but it shouldn't stop there. Often, safety training is a waste of time and money. I hate saying that, but it is true. Having a safety professional read slides to people is not my idea of effective training. Effective training places someone in a position that they may not be comfortable with; it increases the pressure on them to make quick decisions, and it should be designed to humble participants. Only then can you control for complacency and perceived mastery among human beings. Leadership must mandate training of the workforce to control for human risk factors. If they don't, who will?

Leaders protect the workforce. Have you ever worked at a place where there doesn't seem to be much difference between the treatment of those who do what is expected and the treatment of those who do not? This not only kills motivation of the workforce but also places the workforce at an increased risk.

Safety is not an individual action, it is a collective one. For example, if I wear my seatbelt and follow all rules of the road when I am driving, it does not guarantee my safety. The environment, the condition of my car, and how others around me are driving will have direct impacts on my safety in addition to my individual behavior. We may not be able to control the environment or even equipment failures, but we can control the behavior of others by holding them accountable to specific expectations.

Leaders protect the workforce in two ways. First, in order to develop trust—a key ingredient of culture—leaders must accept that humans make mistakes. Even the best intentions may fall below expectation. If this happens, leaders must protect the integrity, reputation, and motivation of those who demonstrate the human characteristic of flaw. I would not consider anyone who demonstrates an effort to comply with expectations a failure, even if they do fail to live up to the expectations. In time, humans can and will live up to expectations if we believe in and adequately nurture them. I believe you hire for attitude and train for skills. But what happens when you must hire for skills and deal with attitude?

The second way leaders protect the workforce is by dealing with negative attitudes. A negative attitude is the rejection of expectations, a challenge to common beliefs and behaviors, and a threat to any existing culture. I am not saying that people should not be able to think or behave for themselves, but if you belong to an organization and the organization has established expectations, you must do your best to live up to them. If you intentionally ignore or disregard the expectations, leadership must intervene to protect the workforce.

Usually leaders don't have problems with the first level of protection. I find that most leaders are quite forgiving and understanding of human error. However, the same forgiveness should not be given to individuals who blatantly disregard the organization's behavioral expectations. Placing oneself over the entire organization is a risk that leadership must own and control. Leaders accept the workforce for being imperfect and protect them from dismissal and scrutiny when an accident occurs. Leaders also protect the workforce by directly challenging those who choose to ignore organizational expectations.

Leaders promote safety. A leader must lead by example and promote safety through his or her words and actions. What happens when leadership contradicts safety with their words or actions? This promotes silos, discourages trust, and certainly destroys the likelihood of a safety culture. Leaders must promote the overall safety program. In doing so, leadership must ensure that they are on the same page with the safety professionals.

Promoting the workforce also means implementing positive and negative accountability strategies in the workforce. Leaders must provide feedback to the workforce about what they are doing correctly and what they are doing incorrectly. This promotion of and attention to safety behaviors provide additional reasoning for safer behaviors to occur. Obviously, if someone is not doing something safely, something needs to change.

When things need to change, leadership works with safety officers to do it together, making safety a top priority. Leadership ensures that everyone in the organization is connected and rallied around safety. When leadership promotes safety, they flatten the organization by serving with and by both the safety professionals and workforce. Solving safety issues requires partnership within an organization at all levels. It requires discussions, training, and review and even promotes disagreements. You don't have to agree with, like, or enjoy those you work around, but you do have to share commonalities in order for a safety culture to exist.

Leaders CARE. Leaders ensure compliance with accountability when resources and expectations are provided: CARE. Leadership can provide expectations and resources. But is that enough? No. If leadership doesn't care whether the workforce complies with expectations, why would the workforce care? Show me that you don't care, and I will show you that I am better than you at not caring!

Do you know how much safety officials can suffer when leaders don't seem to care? Loss of sleep, stress, anxiety, and even depression can occur simply because a leader stops after providing expectations and resources. When all a leader does is say that the rules must be followed, it's the safety official who runs around like crazy, trying to get the workforce to comply with them. Compliance means the workforce is making an effort to live up to expectations. Not everyone who tries will succeed; therefore, we implement positive and negative accountability controls.

Positive accountability means that leadership praises those who are living up to expectation. This is important because it motivates the workforce to follow an example that leadership appreciates. It also sends a strong message that leadership remains engaged well beyond the establishment of expectations and provision of resources. But what happens when the workforce is not living up to expectations?

Negative accountability is not negative at all. "Negative" refers to the job not getting done. Leaders need to provide this accountability in a constructive manner that increases motivation, ability, and the willingness of the workforce to try harder. If the negative accountability is destructive, it includes punishment, which only increases resentment and hides true behaviors from those who are punishing. When leaders care, the workforce does, too.

Workforce Expectations

We have set the stage with leadership, defining expectations that will prepare and protect the workforce while promoting safety. If the workforce is properly prepared and protected, the expectations set by leadership should be accepted within any organization working with an infectious disease. These expectations are behaviors that can be implemented today and can demonstrate profound increases in safety almost immediately. They are based on past lessons that are often ignored, even today, in both laboratory and hospital settings. A lesson will continue to present itself until it is learned. What is the difference between a lesson learned and a lesson ignored? Change.

I have asked scientists and health care workers around the world about their willingness to comply with five behavioral expectations. These expectations are: (1) I will follow all SOPs to the best of my ability; (2) I will ensure that others follow SOPs to the best of their ability; (3) I will report all accidents, incidents, and near misses; (4) I will report any symptoms which match the clinical presentation of the agents I working with (and around); and (5) I will report any new or known medical conditions which may place myself or others I work with at an increased risk.

Most of those I have trained agree with the logic or merit of these five expectations. They realize there is value and importance in each. However, believe it or not, being willing to actually follow and implement these expectations is a true challenge for many!

Follow all SOPs to the best of your ability. The main difference between a safety climate and safety culture is that the workforce actually follows SOPs. The death of Beth Griffin was a tragedy that could have been prevented. At the time Beth was working with macaques, there was an SOP that required those working with animals to wear eye protection. However, nobody in the organization wore eye protection, and there were no consequences for choosing to violate the SOP. Thus, the culture permitted the violation to occur, resulting in a young woman's exposure to herpes B virus that led to her death.

SOPs should be strategic plans, based on risk assessments and proven to be effective at minimizing risks associated with infectious disease. We know a human risk factor is that humans will become comfortable with a risk they work around frequently. Adherence to SOPs controls for this risk and would have made a vital difference in Beth's case.

Ensure that others follow SOPs. Safety is not an individual effort but a collective one. If only one member of a culture is behaving and others working around that individual are not, the risk to the individual and collective group could be substantial. I mention Beth's case here again because I believe these first two expectations establish the core foundation for ensuring shared beliefs and behaviors among people with different perspectives, education, and experience levels, creating a safety culture. When individuals who belong to a unit do not behave consistently with one another, there is no culture of safety.

One of the most important factors for any culture is that it must self-regulate, meaning that the workforce must be able to challenge others when SOPs are not being followed. If you are a leader in an organization reading this today, I suggest you break an SOP in front of witnesses in the workforce. Make sure they see you doing it. If they confront you, praise them. If they don't confront you, ask them if they noticed your behavior, and then empower them to speak up when they see others not following SOPs. Be kind and serious. These first two expectations lay the foundation for a culture of safety within any organization.

Report all incidents, accidents, and near misses. Moving from the biological safety arena into chemical safety, I want you to know about the death of another young woman. At the University of California, Los Angeles, Sheri Sangji died from serious burns as a result of many failures. These included the improper procedure she used to handle chemicals and the fact that she did not wear a laboratory coat; most importantly, previously identified safety issues had been ignored and remained uncorrected at the time of her death. A special aspect of this death is that, for the first time, the laboratory leader was criminally charged: the Los Angeles District Attorney filed four felony charges against the Regents of the University of California and Patrick Harran for "willful violation of safety regulations."

If leadership lives up to the expectation that they will protect the workforce, trust can be established, and when unexpected events occur, they will be reported. One of the main reasons people do not report incidents, accidents, and near misses is because they fear being punished. This could lead to a devastating situation for two reasons. First, if an organization does not act on incidents, accidents, and near misses (from a collective group standpoint), then only one individual improves (the one who makes a mistake or experiences the incident) rather than the overall organization itself. Second, if an incident is not reported, we never know if it leads to illness. This matters because an incident cannot be officially titled an accident (something causing harm) or near miss (something that did not cause harm) until the incubation period of the infectious agent being worked with is past. If an incident occurs, it must be reported so the workforce can be supported with prophylactic countermeasures and surveillance systems, not only to avert disease but especially to reduce the likelihood of the incident being repeated.

Organizations become safer when they are learning and changing, especially following an incident, accident, or near miss. While I was serving in the Emory Isolation Unit, there was a long list of incidents which we reviewed every day with all health care staff to ensure that the lessons were learned rather than ignored. Learning will take place if trust exists between leadership and the workforce.

Report symptoms that match the clinical presentation of agents. In 2004, three individuals working in a BSL2 laboratory at Boston University became sick with pneumonic tularemia. The laboratory was working with an attenuated strain of *Francisella tularensis*. Usually those working with this agent work at BSL3 laboratories and within a biosafety cabinet, because it can be spread through aerosol routes; however, because it was attenuated, the biosafety committee allowed it to be worked with at a lower level. Here comes another consequence of safety climate versus culture.

The safety committee wanted an extra layer of safety redundancy and required the work to be done in a biosafety cabinet. However, because the biosafety cabinets were being used for storage, the request was ignored, and work was done on the open bench. Two members of the workforce became sick with a pneumonic condition in May and another in September. Thankfully, all were treated successfully, but it took three cases of illness before the organization realized that the laboratory was the source of the illness. I have long believed that we are seeing only the tip of the iceberg of illnesses that are acquired in the laboratory. We must do a better job with this!

One of the most negligent things I have witnessed in the workforce is lack of knowledge. Most of those who work in laboratories or isolation units don't know the answer to the question, "How would infection with the agent you are working around present clinically in your body?" It astonishes me that there are animal care workers, laboratorians, cleaners, nurses, and even doctors who cannot answer this question. If they do not know how infection with the agent clinically presents, how would they know if that agent was making them sick?

When I have responded to infectious disease outbreaks worldwide, I have noticed a certain psychological effect. The general public is very concerned about infectious diseases. If threatened with infectious disease, their response is fear, stigma, and then denial. This means that they fear anything to do with the infectious disease, so anyone or anything infected is immediately stigmatized (or rejected), and science is completely denied. I have noticed this with HIV, severe acute respiratory syndrome, and especially the Ebola outbreaks. Any cough, sniffle, or fever is immediately feared, and the overuse of medication or medical treatment is very common during infectious disease outbreaks.

But how do doctors, nurses, and laboratorians respond? I have observed that when you are comfortable around infectious diseases your psychological predisposition is first denial, then fear, and then stigma. If you begin experiencing any symptoms that match the clinical presentation of the agent you are working around, you immediately deny that it could be that agent. After what you thought it was turns out to be wrong, then you begin to fear that it could be the agent you are working with. When this realization occurs, you become embarrassed and fear being stigmatized because you should have known better. The fact is, reporting any clinical symptom that matches one caused by the agents you are working with falls within a medical surveillance program. Medical surveillance monitors the real-time health information of the workforce, ensuring that if the environment is causing the illness, a trend is detected and strategies are put in place to investigate for the protection of the workforce.

A final note: It doesn't matter if the agent you are working with presents with flu-like symptoms, so that an infection can be easily dismissed as "a cold." That is not the point and serves as the greatest excuse of those serving on the frontlines of infectious disease. Safety is not about one person, it is about the group. If multiple people present with illness within an incubation period, it means there could have been a common-source exposure. Maybe it was the lunch room, maybe a staff meeting. Organizations should have systems in place which can detect and investigate situations such as a cluster of flu-like symptoms, instead of ignoring or missing them altogether.

Report any known or new medical conditions which may place myself or others I work with at an increased risk. The deaths of Linda Reese and Malcolm Casadaban are described earlier in this book. In both of these tragedies, there were existing medical conditions that increased their personal risk of working around infectious agents. As time passes and we age, our immune systems fluctuate with both chronic and acute conditions. These fluctuations have direct impacts on our overall health and safety.

It is not a question of whether we have an acute or chronic condition that may impact our overall risk, it is a matter of when. It could be as simple as stress or pregnancy or diabetes. Or it could involve being prescribed a simple steroid or other medication that compromises immune function. Maybe you are caring for an elderly individual or someone fighting cancer. The fact is that your personal health and the health of those you work and live around are connected. If you work with infectious agents, you should notify leadership of any acute or chronic medical conditions that may increase your risk while doing that work.

Summary. Each of these five behavioral expectations is tied to the incidents I have described. These are examples which we know about as a result of a tragic loss. I cannot imagine the lessons we would learn if we knew more, in less severe cases. There is an enormous need for a central reporting warehouse of both laboratory- and hospital-acquired infections. The information gained by a central reporting system would provide valuable information on both workforce and patient safety. It would also challenge our current processes and ensure that they demonstrate the results we want: to protect the workforce and patients against infectious diseases.

If you work with or around infectious agents, consider implementing these five behavioral expectations today. At a minimum, discuss them with the workforce. Consider them proposed solutions from those who have lost their lives serving on the frontline of infectious disease.

Safety Official Expectations

The third perspective, that of the safety official, requires expectations as well. If each perspective follows the expectations presented here, it will serve as the glue for the One-Safe culture with the blending of leadership, workforce, and safety. If you are a safety official, consider the following expectations.

Know the standards of the profession. When it comes to serving safety, it is very important that safety officials know the standards of the profession they serve. I have witnessed safety professionals who take safety as a personal endeavor, making recommendations and claims that extend well beyond the standards of the profession. As explained in chapter 5 under the concept of applied biosafety, the job of a safety professional is to know the standards and make recommendations based on the standards and their professional expertise. Not one or the other; both.

For example, per the existing United States biosafety standards, biosafety cabinets are not required in BSL2 laboratories. Regardless, the safety professional should consider the processes, capability of staff, and available resources when developing an organization's workforce practice specific to biosafety cabinets in BSL2 laboratories. Safety professionals should be able to make sound recommendations based on the standards of the profession while cautioning those they serve to consider additional organizational factors that may not be adequately covered in the existing safety standards.

Serve rather than police. "*Pssst!* Safety officer coming!" All of a sudden, everyone begins to do what they know they are supposed to be doing because the safety officer is around. Does this sound familiar to you? If so, chances are you are working in a safety climate (not culture) and within an organization that punishes rather than serves.

When safety professionals choose to police individuals, it has a very negative impact on behavior. The aim of policing is to see what is not working, what is going wrong, and what needs to be stopped. It does not consider the need to behave, nor does it teach alternative behaviors to satisfy those needs. So rather than *policing*, safety officials should consider *serving*, which removes judgment (right/wrong, good/bad) and fosters the development of the psychological safety net needed to create a safety culture rather than a climate. This is why placing a safety official in a regulatory role is a huge mistake. Safety officials must be seen as people who are serving rather than regulating. If an issue or problem occurs, safety personnel should be positioned as the best friend you call first, rather than the last person you call.

Serving means that you aim to replace old behaviors with new behaviors. It makes older behaviors harder and newer behaviors easier. When a safety professional serves, he or she empathizes with why a behavior is occurring, becomes aware of the implications of that behavior, and manages the process of replacing that behavior both emotionally and behaviorally. All safety professionals should serve and do it with a smile. I know that sounds cheesy, but making yourself smile is a self-regulating exercise that not only makes you feel better but also makes others feel better, too.

Increase capability rather than dependency. In an earlier chapter, I thoroughly discussed intrinsic safety. Safety should not occur just because the safety official is present. Instead, safety must be something that the workforce can do for themselves. To achieve this, safety officials must begin including the workforce in safety activities. These activities include the development of safety plans, facilitation of safety audits, and reporting of safety accidents to leadership when they occur. Safety professionals cannot do safety to and for the workforce; they must do safety *with* them so it can be done *by* them. Safety professionals must increase capacity rather than dependency. They do this by implementing intrinsic safety strategies when serving the workforce.

Be seen. If you are not seen, then you are not there. Safety professionals are the connection between the workforce and leadership. You cannot do safety from behind a desk. You must get out, ask questions, be interested in what is happening, and do your best to solve problems proactively. This means you must walk around, shake hands, share a cup of coffee, and observe. When you see something wrong, don't judge it or focus on it unless it is going to kill or hurt someone. Instead, walk around and focus on what is working and what people are doing right.

If you see a problem, ask about it, learn what is causing it, and be empathetic rather than judgmental. State that you have an issue with the behavior you are witnessing and want to understand it, to see if you can work together to find alternatives to it. As Lee Alderman and Henry Mathews taught me, safety is about checking in every day, being there, and being seen as often as possible. So get out, take a walk, and be seen for safety.

Advocate for all. Safety advocates for the workforce and for leadership. It serves both, not one over the other and certainly not itself solely. In a One-Safe culture, safety must be the keystone that keeps leadership and the workforce together.

Advocating for the workforce means that safety professionals ensure that the workforce is prepared and protected at all times. Advocating for leadership means that the safety professionals ensure that leadership is informed about workforce expectations. Safety officials should never be enforcing expectations. That is the job of leadership. Safety serves to recognize when expectations are being met and when they are not. Safety professionals serve with the workforce to achieve expectations; it is leadership's job to provide accountability when an expectation is or is not met.

CONCLUSION

Each perspective—leadership, workforce, and safety professional—has an important role in the One-Safe culture. Blending doesn't negate that fact; it recognizes it. By providing expectations for each perspective, those in the culture can understand who is doing what and can begin to trust in the collective safety effort.

Safety cultures exist but are often siloed. When siloing occurs, safety offices are located somewhere in the basement or far away from where the work takes place. Leadership provides resources and ensures that expectations are established, but beyond that they disengage, allowing the workforce to do their job. The workforce tolerates both the leadership and safety, understanding that they are legacy roles in the organization. When these three perspectives are siloed, the gaps that exist produce risk to everyone inside and outside the organization.

Providing leadership, safety officials, and the workforce with common risks, rules, and rituals while expecting them to adhere to specific expectations blends perspectives and establishes common beliefs, allowing common behaviors to develop. In that way, we develop a One-Safe culture—a collective effort by all for the safety of all, both inside and outside the organization.

BIOSAFETY *in the First Person*
An Incredible Journey by Robert Hawley

The first time I met Bob Hawley was at the annual conference of the American Biological Safety Association. I asked if I could speak with him for a moment, and he gave me not only a moment, but insights and perspectives on biosafety that I carry with me to this day. As I teach the details of the BMBL in my trainings, it is shocking to many participants to realize the differences between what they think is required and what actually is required. I learned this the hard way, losing a $50 bet to Bob on a specific OSHA (Occupational Safety and Health Administration) regulation. The man not only knows OSHA, he knows the importance of knowing the guidelines that represent the profession you serve.

While working at the U.S. Army Medical Research Institute of Infectious Diseases (USAMRIID) in October 2001, I was afforded the opportunity of a lifetime: to participate in a national biosafety incident and also to share some of my sampling and decontamination experiences and apply lessons learned. I was honored to be asked to provide consulting services to the CDC at the West Palm Beach Emergency Operations Center (EOC), West Palm Beach, Florida, regarding an incident involving the release of *Bacillus anthracis* spores, the causative agent of anthrax, that occurred at American Media, Inc. (AMI), in Boca Raton on October 2, 2001 (1). At the time of the incident, AMI was a publisher of magazines, supermarket tabloids, and books. Throughout my visit, I had the opportunity to interact with many individuals from different U.S. local and national agencies. What an incredible journey and another opportunity to learn!

I primarily provided assistance to individuals from the U.S. Environmental Protection Agency (EPA) in Atlanta, Georgia. The EPA was brought in to take the lead for control, do the decontamination (DECON) procedure(s), and develop alternatives. My objective was to offer a rational approach to the sampling and DECON of the facilities affected. My goals included the consideration of both clinical and environmental sampling data in recommending a DECON procedure; the involvement of the media; the explanation of terminology (DECON, sterilization, disinfection, exposure, infection, and disease); and rationalizing the contrasting procedural approaches for a hazardous materials incident involving industrial chemicals versus an incident involving a biological agent. I discussed the role of USAMRIID assets that were potentially available for sample identification purposes, emphasizing our mission, how our assets might be used, and whom to contact for further questions. The information provided to other agencies during my visit was well received and appreciated, judging from their comments and the quality of their questions and our discussions.

During our discussions at the EOC, all attendees realized that the selected methods of DECON and other procedures done at the AMI building would set a precedent for future national incidents. DECON options included using soybean emulsion, a foam developed by Sandia National Laboratories, formaldehyde vapor, and household bleach. In the context of the action of these decontaminating agents, I discussed the differences between sterilization (complete destruction of living microorganisms) and disinfection (selective elimination of certain undesirable microorganisms in order to prevent their transmission) and between exposure (introduction to a microorganism) and infection (introduction of a microorganism into the body with a subsequent response). I also emphasized that DECON is disinfection or sterilization of infected articles in order to make them suitable for use. I emphasized the distinction between absolute safety, which would require sterilization of the affected areas, and reasonable safety, which could be achieved through a decontamination or disinfection process, with the latter entailing some low level of risk.

I spoke about the importance of involving the media, as a consequence of their presence at the EOC one evening, and of keeping them informed of developments, with the understanding that some information must be carefully phrased or withheld because of the continuing investigation. I mentioned that the media could be used to our advantage because they would relate this information to the community, which would foster and enhance credibility.

A member of the EPA mentioned during the morning of October 15 that a cleanup of affected sites at the Boca Raton post office would be done during the evening. I worked with the EPA Decontamination Team for a few hours developing a sampling and DECON plan. I explained the procedures for surface DECON of equipment, such as computers and file cabinets. We would use a 0.5% sodium hypochlorite solution, which is a 1:10 dilution of commercially available liquid household bleach (5.25%) containing about 5,500 parts per million (ppm) of freely available chlorine. After 5 minutes of contact time, this produces a greater than 90% inactivation of *B. anthracis* spores (up to a 3 $\log_{10}$ reduction in viability), per published reports (2–5).

According to procedure, personnel doing the DECON applied the 0.5% sodium hypochlorite solution to affected nonporous surfaces, such as work surfaces, computer equipment, file cabinets, vinyl floors, painted walls, and ceilings, for a minimum contact time of 5 minutes. During DECON, individuals wore personal protective equipment (PPE) to protect them from a splash. PPE included a Tyvek coverall to protect their street clothes, shoe covers to protect their shoes, a minimum of a fitted N95 half-face disposable respirator to protect against nuisance dusts, eye protection (wraparound goggles or full-face shield), and two pairs of surgical (latex or nitrile) gloves.

Following the contact time, the decontaminated areas were rinsed (wiped) twice with a water-moistened cloth or sponge to remove any residual sodium hypochlorite solution. Samples were obtained after the DECON process to validate efficacy of the procedure. Upon completion of the DECON, cleanup materials (liquid sodium hypochlorite, moistened cloths, and sponges) were disposed of using the normal household waste stream. At the point of exit from the decontaminated area, personnel removed their PPE in reverse order of donning, removing the outer pair of surgical gloves, Tyvek coverall, shoe covers, eye protection, and mask. The last item to be removed was the second pair of surgical gloves. Personnel then immediately washed their hands with soap and warm water for at least 30 seconds.

For DECON of porous surfaces, such as carpeting and cloth-covered furniture, individuals wore PPE as previously described for nonporous surfaces. Areas or materials identified as contaminated were decontaminated in place by application of 0.5% sodium hypochlorite. Treated material(s) would be removed and appropriately disposed of and the material(s) replaced as required. Carpeted and/or cloth-covered areas or materials identified as not contaminated were vacuumed with a high-efficiency particulate air (HEPA)-filtered vacuum to minimize airborne allergenic particulates and help provide better indoor air quality. At the point of exit from the decontaminated area, personnel followed the procedures as described for nonporous surfaces.

I participated in the pre-DECON sampling and DECON of nonporous areas of the Boca Raton main post office facility with two individuals from the EPA. I prepared the tubes for sampling (moistening swabs with distilled water and numbering sampling tubes) and prepared 5 gallons of the 0.5% sodium hypochlorite solution for DECON and the two 5-gallon water wash containers. I directed sampling to include each of the four sides of the rectangular mail slots (each approximately 6 by 6 by 10 inches) and each of the four corners of each slot. The sample swab was carefully returned to the BBL Culturette container without crushing the glass containing the sample preservative. A member of the EPA team and I then traveled to the West Palm Beach County Health Department in Lantana, Florida, to deliver the 11 pre-DECON samples to the USAMRIID team who were already on site. My assistance visit ended October 16, 2001, and I returned to USAMRIID.

The on-site experience I gained was professionally rewarding, as I was given the time to share my experiences with colleagues from the CDC and EPA and obtain new information on sampling and decontamination strategies. I also gleaned much information on the dynamics of group interactions and how "talking things through" promoted the teamwork approach. The experience I gained was the foundation of an incredible journey.

REFERENCES

1. **Traeger MS, Wiersma ST, Rosenstein NE, Malecki JM, Shepard CW, Raghunathan PL, Pillai SP, Popovic T, Sejvar JJ, Dull PM, Tierney BC, Jones JD, Perkins BA, the Florida Investigation Team.** 2002. First case of bioterrorism-related inhalational anthrax in the United States, Palm Beach County, Florida, 2001. *Emerg Infect Dis* **8:**1029–1034. https://wwwnc.cdc.gov/eid/article/8/10/pdfs/02–0354.pdf

2. **Russell AD.** 1990. Bacterial spores and chemical sporicidal agents. *Clin Microbiol Rev* **3:**99–119.

3. Sagripanti J-L, Bonifacino A. 1996. Comparative sporicidal effects of liquid chemical agents. *Appl Environ Microbiol* **62:**545–551.

4. Brazis AR, Leslie JE, Kabler PW, Woodward RL. 1958. The inactivation of spores of *Bacillus globigii* and *Bacillus anthracis* by free available chlorine. *Appl Microbiol* **6:**338–342.

5. Hawley RJ, Eitzen EM Jr. 2001. Biological weapons—a primer for microbiologists. *Annu Rev Microbiol* **55:**235–253. http://journals.sagepub.com/doi/pdf/10.1177/153567600100600303

chapter 11

Emergency Preparedness and Response to Biological Risks

I was 10 years old, and my brother was eight. My parents were leaving us alone for the very first time, allowing me to babysit my brother, but in reality, it was more about both of us helping each other out. Before leaving, my parents walked through a strict set of instructions about what to do if a stranger came to the door after they had left. They went through all scenarios and set expectations for us. This happened over several days until the day finally arrived. I remember being excited. I was going to cook dinner and make sure everything was done according to plan. It was an exciting chance to prove myself and be free from my mom and dad for a couple of hours.

My parents went over everything again and then said good-bye. We were free. Hungry Man chicken pot pies were in the oven, cartoons were on the TV, and we were enjoying life. What happened next is something I still remember vividly 35 years later.

There was a knock on the door. My brother and I looked at each other. This was unexpected, and when we looked through the side window to see who it was, without being seen, we did not know the person. We were watching the person as he continued to knock, nervous but hopeful he would go away, and he did. To this point, we had done everything according to plan. We had not said anything or been seen. My brother and I both sighed in relief until he returned two minutes later. At that point, he knocked harder and then began attempting to pull the screen door open. My brother and I immediately turned around to run out of the back of the house, and right there were our parents watching us.

It took some hugs and kisses to calm us down, but we had not been hurt, my parents had tested the plan, and my brother and I received our first training in emergency response. The next time a stranger knocked on the door, we were more prepared. During the next 10 years of my life, our parents repeated similar exercises with choking, drowning, fire alarms, and military deployments. How my parents trained me has had a direct influence on how I train others and how I have responded in emergency situations.

In 2014, shortly after the 10-year NIH training grant expired, I was working part-time at Emory University while consulting full-time with our small business. I received a call and was asked to visit the isolation unit where the first two cases of Ebola in the United States would be treated. My visit had a goal: to observe the staff assigned to work in the isolation unit and see if they were prepared to do so. I have learned through many emergency response situations that it is not what people do when things are optimal that determines preparedness, it is what they do when things are less than optimal. I asked the staff three questions. (1) What would you do if the patient had an event of uncontrolled bloody diarrhea or vomiting? (2) How would you doff your personal protective equipment (PPE) once it was completely contaminated from this event? (3) If you had to evacuate a latent-stage Ebola patient because of an emergency, how would you do it? As I listened to their answers, it became very clear that the staff, who believed they were ready, were not ready at all.

The basic definition of an emergency is an unexpected event that exceeds existing resources. This requires being prepared to respond. However, don't make the mistake that many organizations do by stopping there. Although many invest heavily in preparedness, it is recovery and mitigation that typically save the greatest amount of resources while minimizing losses following an emergency situation. Emergencies are going to happen even when you do your best to prepare for them. Some experts believe it is what you do *before* the emergency that matters the most in terms of outcome. However, because humans naturally do what they can to survive, what we do *following* an emergency can matter more. Regardless, all professionals working around infectious disease must invest in emergency preparedness, response, recovery, and mitigation.

PREPAREDNESS

The preparedness phase occurs before the emergency and is based on lessons learned, such as those collected during the mitigation phase of a prior emergency. Preparedness involves attempts to secure needed resources, provide strategic plans, establish needed partnerships, and develop proven strategies to respond to the emergency. In my years of working in emergency response specific to infectious disease, I have learned that there is only so much that we can prepare for. We will always have "canaries," those who are at the front of the emergency who signal that a problem exists. Depending on where the problem is, what it is, and whom it is impacting, it can elicit widespread panic, hoarding of resources, vicarious rehearsal, and societal interruption. The preparedness phase occurs until there is an immediate risk to health and safety.

RESPONSE

The response phase happens once a risk is present and is posing a threat. There is no time better than an emergency to see that a plan is an outcome minus the behavior. We can spend years preparing for an event, only to throw out the response plan immediately following the emergency because we did not comprehend the extent of the emergency during the preparedness phase.

During the response phase of any biological risk, you must provide behaviors for people, including the public, to do while determining what exactly is happening. During the 2001 anthrax attacks, public health messaging could not keep up with the press, resulting in mixed messages about safe behavior. This happened again in 2003, with the severe acute respiratory syndrome outbreak. In 2004, the CDC decided that any time an infectious disease outbreak occurred, they would immediately release instructions to the public to follow these generic behaviors:

1. Wash and sanitize your hands frequently.
2. Stay active and exercise.
3. Practice social distancing.
4. Stay hydrated.
5. If you are sick, stay home.

These public health recommendations can be applied to all infectious diseases. While scientists are attempting to determine the specifics of the outbreak, health care providers and the general public can begin responding in an appropriate manner.

No matter how much preparedness you have, understand that you are making decisions and creating a response plan based on limited information. If you wait for all the information to come in during an emergency, you will likely fail. If you make decisions based on limited information, you may fail. Sometimes during emergency situations we fail. It's important to detail and justify every decision you make during an emergency to ensure that history does not distort facts. History can be brutal, and hindsight is always 20/20. Exceptional leaders have made great decisions that failed, and history judged them as though they had all the information when they made the decisions.

RECOVERY

The recovery phase occurs when the immediate risk to health and safety has been mitigated. Notice that I wrote "immediate." Humans may survive, but what is the "new normal" afterward? There are still people who are dying from cancer related to the emergency response efforts of September 11, 2001. Although the response phase is usually over very quickly, the recovery phase may take much longer because of lasting impacts.

What you do during the response phase matters, but what you do following the response phase matters most. A good recovery phase consists of 50% empathy, 25% information, and 25% commitment. An organization must be able to empathize with those who have been directly infected or affected by the biological emergency. They must be able to provide information about what they do and don't know to those impacted. Finally, they must demonstrate a continued commitment to engage, communicate, and serve those impacted. Recovery often comes with blame. You know you have entered the recovery stage when the fingers start pointing. We can minimize the risks of litigation by structuring a recovery phase that focuses on empathy, service, and commitment to those who have been affected. Don't stop when the obvious risk to health and safety ceases; continue until the emotional impact begins to diminish as well.

MITIGATION

The mitigation phase usually begins during the end stages of the recovery phase. The purpose of the mitigation phase is to capture what worked so we can repeat those actions in future response and recovery efforts. In addition, the mitigation phase explores what didn't work, so we can replace those efforts and ensure that failures do not occur again in the response and recovery phases.

The difference between a lesson learned and a lesson ignored is change. The mitigation phase is one of the most overlooked phases because nobody wants to discuss what went wrong, what didn't work, and what could have been done better. However, time has a way of burying those lessons, leading us to repeat behaviors that cause problems. Following any emergency situation, all organizations should review what worked and what didn't work. Leaders and frontline staff should partner to capture these lessons. The simple exercise of asking each person what he or she would start doing, would stop doing, and would continue doing can capture extremely valuable information for the mitigation phase.

APPLIED LABORATORY EMERGENCY RESPONSE TRAINING

Apply the four phases of an emergency to even small incidents like needlesticks or spills. They are even more critical in emergencies that are not one dimensional, where several organizations must respond at the same time. Consider a hostage situation in a hospital or laboratory setting, which requires partnerships among police, fire, and emergency medical services (EMS). How do we prepare for this kind of emergency situation?

Most people will do whatever they can to avoid infectious disease—taking vitamins, eating healthfully, and getting vaccinations. To most people, those on the frontlines of infectious disease who run toward infectious diseases instead of away from them are not normal. Infectious diseases are scary even to many of the emergency responders we would need during a hostage situation, for example. If the emergency responders don't understand the biological risks that exist in a specific emergency situation, how effectively are they going to respond? This is why it is extremely important to partner with emergency responders to develop, test, and evaluate emergency response plans involving biological risks.

The importance of this was driven home in 2006 when I was facilitating a BSL4 training. As I was watching an emergency response drill, something did not feel right. Working at Emory University, I had access to one of the best level 1 trauma centers, so I asked for help (more about this in "Unconscious Individuals," below.) This is when I met James Augustine. He worked with Lee Alderman, Henry Mathews, and myself to develop the Applied Laboratory Emergency Response Training (ALERT) program, which aims to blend laboratory emergency response training with local emergency response training. Since then, we have trained more than 3,000 emergency responders with this program and facilitated tactical exercises ranging from workplace violence and hostage situations to bombs, earthquakes, and many other emergencies. We learned that emergency responders in general have a great level of anxiety about working around infectious diseases. We also learned that those who transfer patients place themselves at enormous risk when encountering an emerging infectious disease. We did not know the impacts of this program until we visited Boston University in 2008.

In 2008, a BSL4 laboratory had been built by the Boston University and the National Institutes of Health. When I toured the BSL4 facility, it was my opinion that it was one of the best-built BSL4 laboratories in the world. It has a great scientific team. However, in 2004 there had been an event where three scientists got sick with tularemia, a public relations nightmare. This emergency led the Boston Public Health Commission to regulate all BSL3 and BSL4 laboratories within the city limits of Boston. This is why our team was invited to facilitate the ALERT program for emergency responders in Boston who would be expected to respond if something went wrong at the BSL4 laboratory. More than 300 participants came from a wide variety of agencies, including police, fire, EMS, public health, building department, and transportation. The breakdown of the participants' affiliations is shown below.

- 49% police
- 25% fire department
- 10% EMS
- 7% public health
- 5% building department
- 1% transportation
- 3% other

Boston University facilitated the 3-hour ALERT program inside the new laboratory, allowing emergency responders to get more comfortable in the laboratory setting. During awareness training, participants rotated among three stations each hour. At the beginning, participants completed a pretraining assessment and pretraining risk perception evaluation, and after finishing the program, they completed the same assessment and evaluation. This allowed us to measure the impact of the program on emergency responder knowledge and risk perception factors related to biological

Table 11.1 Assessment[a]

1. Directional airflow keeps pathogens from leaving BSL3 and BSL4 laboratories.
2. Biosafety cabinets are where pathogens are grown.
3. Every BSL4 laboratory is mandated to have two separate HVAC systems.
4. Halon gas fire-suppression systems are used in most BSL3 laboratories.
5. FBI background checks are required for all staff working in BSL4 laboratories.
6. If someone is exposed to a pathogen, he or she will get sick.
7. The four primary controls for a laboratory environment are administrative, PPE, SOPs, and security.
8. The safety director is responsible for HVAC systems in a BSL4 laboratory.
9. If someone is unconscious, your first step is to check for breathing.
10. During a modified emergency evacuation of a BSL4 laboratory, containment of pathogens is sustained.
11. A power failure at a BSL3 laboratory requires major engagement of emergency responders.
12. Systems failures in BSL3 laboratories are low-probability events.
13. During an emergency, it is a good idea to breach an incubator.
14. When life is at risk, it is important to decontaminate yourself before leaving the laboratory.
15. BSL3 laboratories have steel-gasketed doors.

[a]The pretraining and posttraining assessments are identical. Participants are asked to respond with "true," "false," or "unsure" to each statement.

laboratories. Participants were asked to rate the overall program. The assessment and risk perception evaluation questions are in Tables 11.1 and 11.2. What we learned performing this training not only solidified the ALERT program, it then allowed us to train emergency responders globally, specifically with regard to emerging infectious diseases.

The results we received were quite profound (Tables 11.3 and 11.4). I was most surprised at how well the messages were received. The ALERT program increased awareness and modified risk perceptions associated with biological laboratories. Since 2008, we have replicated this program globally and obtained similar results.

As demonstrated by these results from our ALERT program, emergency responders have much to learn with regard to biological risks and emerging infectious diseases. We found that emergency responders are eager and

Table 11.2 Risk perception evaluation[a]

1. BSL4 risk to general public.
2. BSL4 risk to staff working in the laboratory.
3. BSL4 risk to staff working outside the laboratory.
4. BSL4 risk to emergency responders.
5. BSL4 risk to you.

[a]The pretraining and posttraining evaluations are identical. Participants are asked to rate each risk as high (5), moderate (3), or low (1).

Table 11.3 Assessment

| Statement | Percentage of responses | | | | | |
| | Pretraining | | | Posttraining | | |
	True	False	Unsure	True	False	Unsure
Directional airflow keeps pathogens from leaving BSL3 and BSL4 laboratories.	57	9	34	98	2	0
Biosafety cabinets are where pathogens are grown.	20	27	53	11	89	0
Every BSL4 laboratory is mandated to have two separate HVAC systems.	50	6	44	93	5	2
Halon gas fire suppression systems are used in most BSL3 laboratories.	28	18	54	20	75	4
FBI background checks are required for all staff working in BSL4 laboratories.	68	3	30	100	0	0
If someone is exposed to a pathogen, he or she will get sick.	21	41	38	9	89	3
The four primary controls for a laboratory environment are administrative, PPE, SOPs, and security.	39	8	54	38	62	0
The safety director is responsible for HVAC systems in a BSL4 laboratory.	26	19	54	21	77	2
If someone is unconscious, your first step is to check for breathing.	53	26	22	19	78	3
During a modified emergency evacuation of a BSL4 laboratory, containment of pathogens is sustained.	32	11	57	86	9	5
A power failure at a BSL3 laboratory requires major engagement of emergency responders.	20	32	48	10	86	3
Systems failures in BSL3 laboratories are low-probability events.	36	13	51	39	59	2
During an emergency, it is a good idea to breach an incubator.	7	54	39	8	88	4
When life is at risk, it is important to decontaminate yourself before leaving the laboratory.	53	20	28	28	69	2
BSL3 laboratories have steel-gasketed doors.	22	7	72	25	72	3

willing to be trained to respond properly to these situations. Emergency responders are also eager and willing to assist in developing programs to prepare for and respond to biological risks. It is imperative to prepare and train with those who will be asked to respond to an emergency within a laboratory, hospital, or emerging disease outbreak.

EMERGENCY PREPAREDNESS AND RESPONSE TO BIOLOGICAL RISKS

I believe that when an emergency happens, people should be responding rather than reading. If you are busy reading about what you are supposed to do, how can you be doing what you need to be doing during an emergency

Table 11.4 Risk perception evaluation

Question	Percentage of responses									
	Pretraining					Posttraining				
	High	4	Moderate	2	Low	High	4	Moderate	2	Low
BSL4 risk to general public	17	7	26	14	36	2	1	14	12	71
BSL4 risk to staff working in the laboratory	31	14	31	11	12	16	4	18	27	35
BSL4 risk to staff working outside the laboratory	16	8	28	22	26	2	3	15	12	68
BSL4 risk to emergency responders	32	20	23	10	14	9	3	25	19	43
BSL4 risk to you	22	12	34	11	21	5	3	17	13	62

situation? I will focus on the most important emergency response topics for laboratory and health care staff facing biological risks and emerging infectious diseases. At a bare minimum, you should be trained to respond to the emergencies that are specific to your work setting.

Evacuations

On August 23, 2011, my wife Jacqueline and I were leading an emergency response training program in the BSL3 laboratory of George Mason University in Manassas, Virginia. Before the lunch break, we were discussing emergency evacuation procedures for their high-containment laboratory. In preparation for this training program, I had done a geographical threat assessment, which is a 100-year review of natural disasters in that area. I told them the lab was located near a fault line, so earthquakes were a potential threat requiring emergency response plans. I am not kidding when I say the attendees laughed and joked about having earthquake plans, because none of them had experienced an earthquake, especially around Washington, DC. During our lunch break at a nearby restaurant, I noticed the table shaking. Jacqueline has a tendency to shake her leg when nervous, and I thought this was probably her. But when I looked at her, she said, "It isn't me." It was an earthquake. The dishes in the kitchen were falling. A 5.8 magnitude earthquake was happening, centered in Mineral, Virginia, about 75 miles away. I had the attendees' close attention for the remainder of the afternoon!

We must prepare for all types of emergency evacuation procedures, even if the possibility seems remote. I received a call one day from Chief Paul Burke, who was responsible for preparing emergency responders in the Boston Fire Department for situations that might occur within the BSL4 laboratories at the Boston University National Emerging Infectious Diseases Laboratory (NEIDL). The conversation went something like this:

"Sean, can you believe staff at the NEIDL just told me they would not evacuate a BSL4 laboratory if the fire alarm went off?"

"Chief Burke, no offense, but I wouldn't either. Fire alarms are not supposed to increase risk; they decrease risk. If they are working with an Ebola-infected monkey and run outside because someone overcooked popcorn, this places everyone at an increased risk."

"Sean, if a fire alarm goes off and you are on the 103rd story, am I asking you to jump out of the window to evacuate?"

Chief Burke went on to explain that his goal was not an immediate evacuation but for the scientists in the laboratory to ensure they are taking steps to minimize risks. We worked with Chief Burke, using his insights, to develop a three-tiered evacuation system for the Boston Fire Department. The following describes the green, yellow, and red evacuation procedures.

Green evacuation. A fire alarm goes off. There is no smoke or sign of immediate threat. You are working with a patient, facilitating research with a biological agent, or processing a patient sample. What do you do?

Green evacuation starts with communication. Individuals make a phone call to a predetermined contact who has the responsibility of communicating the situation to those working with infected patients or biological agents. This is a lesson learned in Winnipeg, Canada, after the BSL4 laboratory opened and the organization experienced a kitchen fire, leading to the complete evacuation of the building. Those who were working in BSL4 had no clue what was happening, were abandoned and uninformed, and were not able to make decisions about risk. In short, when an alarm goes off, a contact person should be available to provide steps for what should be done next.

Green evacuations are for situations that are not immediately life threatening. Individuals are asked to secure the biological agents or patients they are working with, follow normal procedures for removing PPE, exit the facility, and, when appropriate, log and report the incident. Green evacuations ensure that biological agents (sources) are secured and risks are contained. Containment is maintained because of the normal doffing of PPE and following of existing procedures.

Yellow evacuation. Someone goes unconscious and is not responsive. You are not a medical professional, and EMS are not coming into the environment to treat the individual. What do you do?

First and foremost, we want to ensure containment to prevent what we are working with from traveling from where we are working with it. Therefore, it is important to call out to notify others of the emergency and request assistance in moving the individual to an emergency exit, removing their PPE in an appropriate order, and moving the individual into a clean corridor. Once the unconscious individual is there, emergency responders can assess and treat the person. Yellow evacuations prepare for situations where containment must be maintained and doffing and exiting procedures are modified to respond to the emergency.

Another example where a yellow evacuation is appropriate might be an earthquake, where no structural damage to the building is noted but equipment may be blocking the main exit. Staff should be trained to doff PPE and exit using modified procedures while ensuring containment of the biological agents they are working with or around.

Red evacuation. Is containment more important than life? Never, for several reasons. Exposures do not always lead to infections. If you are exposed, it takes time to incubate, become sick, and transmit the infection. Most of the time, actions can be taken to make your exposed body a less susceptible host for that biological agent. Leaders should always empower staff to get out if life is at risk. If someone threatens you with a knife, an earthquake causes visible structural damage to your building, or a monkey escapes and you feel your life is at risk, *get out immediately any way you can*.

However, staff are not empowered to act irresponsibly. A red evacuation means there is a loss of containment, and we need to train staff to respond appropriately. Red evacuations require designated routes for exiting and decontamination procedures. I will never forget a lesson learned during an emergency simulation drill at the University of California, Los Angeles.

We simulated a red evacuation specific to an earthquake. We asked scientists who were in Tyvek suits, double gloves, and booties and wearing MaxAir hooded powered, air-purifying respirators to evacuate using the red evacuation procedures. They did an outstanding job, exiting in a manner that did not place those around them at increased risk. As they assembled in the pre-identified meeting location, we waited. The fire chief then asked me a question. "Sean, what are you waiting for?" I said, "Your emergency responders. You are coming to decontaminate us, right?" "Sean, if there is an earthquake that shakes the laboratory so hard that structural damage occurs leading to a red evacuation, you guys will be the last ones we come to!" Because of this exercise, we learned that sometimes a red evacuation will require us to decontaminate ourselves. We assembled a decontamination backpack that includes tarps, gloves, disinfectants, biohazard bags, zip ties, and laminated copies of the procedures.

Is your organization ready for a red evacuation right now that will protect the public from a breach in the containment of biological agents? What if you had an infectious patient and needed to evacuate him or her because of a fire or bomb threat? We must be prepared to respond to all evacuation scenarios, especially when we are working with biological agents or infectious patients. For this, a single evacuation strategy is insufficient. Emergencies may provide time to respond or require an immediate response. These considerations must be included in the preparedness, response, and recovery phases of any emergency preparedness and response plan.

Needlesticks and Eye Splashes

When it comes to needlesticks, there is a gold standard response that should be taught through the professions of biosafety and infection control. Laboratory staff and health care workers should be trained and empowered to respond immediately when their skin is punctured, whether by a needle, scalpel, animal bite, or other type of sharp.

I have seen safety protocols that ask staff to report a needlestick before responding to one! Safety protocols are supposed to reduce risk, not increase it. When the response to a needlestick is delayed in order to report the accident to another person, the chances increase that the exposure will lead to an infection. The goal of this response is to minimize the quantity of the exposure as soon as possible, thereby decreasing the likelihood that illness will occur. One of the few things that all infectious pathogens have in common is that the more you are exposed to them, the greater the likelihood of illness. Exposure does not always lead to infection, but the primary goal, before anything else, is to minimize the exposure.

This requires you to expose the wound immediately, then begin flushing the wound with water. This dilutes the exposure. There is a debate about whether one should express the wound and encourage it to bleed. I can find evidence to support both, but unless the exposure is to a toxin, such as a spider or snake bite, I encourage bleeding at the site of the wound, as this is your body's natural way of protecting itself when the skin is broken.

I have had conversations with several infectious disease doctors practicing medicine around the world about how to expose the wound. I have yet to find one who tells me they would flush a wound for longer than 5 minutes, whether dealing with a biological exposure or a bite from a confirmed rabid dog. So I recommend a 5-minute flush following a needlestick. If you have a splash in your eye, a 5-minute flush will do as well.

However, there seems to be a 15-minute guideline for flushing eyes in laboratories. I can defend a 15-minute flush following a chemical exposure; even though not all chemicals require this, it sometimes takes a 15-minute flush to dilute the chemical enough to stop the damage to the skin. However, have you ever washed your hands for 15 minutes? Think about the potential eye damage or damage to the skin from flushing for 15 minutes. Skin gets waterlogged and becomes weakened, which increases one's risk rather than decreasing it. Five-minute flushes are sufficient enough following a biological exposure—period. When I have questioned the 15-minute requirement, often a safety official will say to me, "We say 15 minutes, so we can get them to flush for 5 minutes." Lying to people to get them to do what is safe for them is not a good practice. Train them to follow the appropriate protocols needed to protect them following an exposure, including flushing for 5 minutes.

After exposing the wound and flushing for 5 minutes to express it, cover the wound to minimize the likelihood of further exposures as PPE is doffed. After covering the wound, doff PPE normally. This kind of emergency requires immediate action but not at the risk of breaching the containment of PPE. The next step is to report for medical assessment, as some agents may require immediate prophylactic treatment following an exposure.

The last step is to always log and report the incident. Failure always provides a lesson. Remember, the difference between a lesson learned and one not learned is change. Following any failure, change should be inevitable. If change doesn't occur, the lesson could repeat itself in the future.

To summarize, the process I endorse for responding to needlesticks is as follows.

1. Expose the wound.
2. Flush and express the wound with water for 5 minutes.
3. Cover the wound.
4. Doff PPE normally.
5. Report for medical assessment.
6. Log and report the incident.

This plan is quick, efficient, easy to learn and remember, and effective. It enables laboratory and health care staff to respond immediately following an exposure rather than waiting for permission or guidance to do so.

Spills

Kent Brantly, the first patient with Ebola to be treated in the United States, and his wife Amber relate the following story in their book *Called for Life*, so I am not breaking patient confidentiality by sharing. Spills occur in laboratories and also occur in hospitals. During Dr. Brantly's treatment at Emory University Hospital, we had such an incident. Dr. Brantly had a bout of uncontrolled diarrhea that was not contained and went throughout the isolation unit. Although in his book he wrote that I was guiding safety efforts from outside the unit, this was the one time that I entered the isolation unit to assist, after watching two nurses freeze.

After traveling to and serving many organizations around the world, I can tell you there are many ways to clean a spill responsibly. However, there are five critical responses in any spill cleanup that should be in all spill standard operating procedures (SOPs). Review your spill cleanup processes; if any one of these five responses is missing, consider correcting your spill SOP.

(i) When a spill occurs, the first action should be to notify those working around you. This not only puts them on alert, it provides aid to you if needed and ensures that they don't come into the area where the spill has occurred. Once those around have been notified, (ii) prevent the spill from being tracked around the workplace. If you can remove your PPE (booties, gloves) you should do so, just outside the outer limits of where the spill has occurred. Again, the action here is to behave in a manner that doesn't spread the spill outside the spill zone.

(iii) The next action is to always address a spill from the outside in. I have seen many make the mistake of cleaning the spill in the direction of

Table 11.5 Ingredients of a biological spill kit

Checklist
Gloves/booties (3)
Tongs
Biohazard bags (4)
Beach towels (3)
Gallon of bleach/MicroChem
Swiffer sweeper (small)
Bleach wipes (1 container)
Zip ties (4)

inside to out, which leads to further contamination of PPE and tracking of the spill throughout the environment.

(iv) The next step sounds easy, but it may prove to be the most difficult of all in a spill cleanup. Ensure proper contact time with the disinfectant on the agent. The reason this is the toughest step is because I have watched as people behave around spills. They spray hundreds of paper towels with disinfectant and set them on the ground like they are assembling a puzzle. Is this as effective as they believe it to be? For example, does the disinfectant soak through the towel and make contact with the contaminated area? Instead, there is a simple solution that is effective. Dip a large beach towel into a bucket of disinfectant and cover the area. This one action is easy, quick, and very effective in making sure surfaces have direct contact with disinfectant for the time needed to inactivate the specific infectious agent.

(v) The final step is to log and report the incident, so there is the opportunity to learn by focusing on *what* went wrong, rather than *who* went wrong. Humans make mistakes. Focusing on what went wrong allows the problem to be fixed, rather than the person.

To summarize, the five critical elements that should be included in all spill emergency response plans are as follows.

1. Notify others.
2. Change PPE to prevent tracking the spill throughout environment.
3. Clean the spill from the outside in.
4. Allow appropriate contact time with disinfectant.
5. Log and report the incident.

Note that most spill kits available for purchase are designed to respond to chemical spills. So build your own spill kit for biological risks. I have provided the list given in Table 11.5 to many laboratories and health care facilities (Table 11.5); in fact, this is the list that was used to assemble the kits for use within the isolation unit before Dr. Brantly's diarrhea spill. The biological spill kits allowed us to respond both quickly and effectively by following an SOP (Table 11.6).

Table 11.6 Sample spill clean-up SOP

1. Immediately alert coworkers.
2. Get spill kit.
3. Remove booties and gloves. Don new PPE.
4. Hang spill sign.
5. Establish spill parameters.
6. Soak towels with appropriate disinfectant.
7. Working from the outside in, cover spill with towels.
8. Remove booties. Don new PPE.
9. Allow appropriate contact time.
10. Dispose of towels and waste in biohazards bag.
11. Mop spill area.
12. Remove booties and gloves. Don new PPE.
13. Log and report the incident.

Gross Contamination

It has been my experience that many organizations don't have emergency response plans for gross contamination. Alternatively, they exist, but they don't work; no organization should ever write SOPs and provide them to the workforce unless they have been tested and demonstrated to work. Proper use of PPE is the method used to reduce risk from gross contamination, so proper doffing usually requires a yellow, or modified, doffing process. Look online at the many ways glove removal is taught, and then test the processes yourself using chocolate sauce or GloGerm. You will clearly see failures in these processes. Failures lead to exposures. Consider the two nurses, Nina Pham and Amber Vinson, who worked in Texas with an Ebola-infected patient. Although they had the best PPE, somehow at some point they were exposed and became infected. I support and teach the "beaking" method of glove removal (see chapter 14).

All doffing SOPs should be tested as though PPE is grossly contaminated. If your doffing SOP doesn't allow staff members to remain clean, the procedure should be replicated and changed until it yields that result. Table 11.7 shows an example of an SOP for doffing PPE at the highest level of BSL3 laboratories or hospital isolation units.

Some will debate the merits of this SOP, especially steps 6 to 9, because they depend on the environment, the agent, and what the person has been doing while working in the area of biological risk. However, when performed correctly, this process has demonstrated a consistently clean outcome. Whatever SOP you approve, ensure that it is evaluated, validated, and verified.

Unconscious Individuals

Earlier in the chapter, I mentioned how a specific emergency response just did not seem right. I am happy that I followed up on that instinct and asked for help. We invited emergency response medical doctors to come to

Table 11.7 Proper doffing of grossly contaminated PPE

1. Remove booties.
2. Remove outer gloves.
3. Sanitize inner gloves.
4. Remove Tyvek suit inside-out.
5. Sanitize gloves.
6. Remove visor.
7. Don new gloves.
8. Move into anteroom.
9. Doff belt, battery, and motor and sanitize respiration system.
10. Remove gloves.
11. Wash hands.
12. Enter shower room.
13. Remove all clothing.
14. Take a 3- to 5-minute body wash. Dry.
15. Enter locker room.
16. Don street clothes.
17. Don jewelry.
18. Swipe out.
19. Exit locker room.

the BSL4 training center to watch how we were responding to an emergency involving an unconscious person. During that response, we opened the individual's BSL4 suit and performed cardiopulmonary resuscitation (CPR) in the laboratory with a breathing bag. The plan was to do this until EMS arrived. Remember, this was in 2006, when biosafety training had just begun for those who would work in BSL4 laboratories today.

The doctors ended up telling us that, in short, our procedures were more likely to kill that individual than save him. They questioned everything, and those questions led to the SOP we have used to train thousands of scientists since. If someone falls unconscious while in a laboratory environment, we must fight our natural instinct to approach and help. Laboratory environments are confined spaces, and trying to help without knowing what caused that person to become unconscious can risk disabling both of you, and quite possibly nobody outside the lab would know about it. So the first step is to call out to communicate to others that someone is down.

The next step is determined by the resources you have available for your emergency response. If you have an automated external defibrillator (AED), make sure everyone in the laboratory or clinic is trained to use it. Generally, the first step is to check for breathing. You can do this by pushing the face shield against the nose and mouth. If the individual is breathing, he or she has a heartbeat and does not need defibrillation. In this case, evacuate the individual, decontaminate, and then wait for medical services to arrive. If

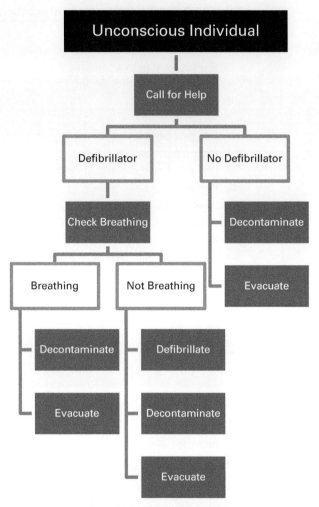

Figure 11.1 Protocol for removing an unconscious individual from a containment area

the individual is not breathing, immediately apply defibrillation, because you are racing against time. Brain damage begins to occur when the heart stops and blood flow ceases, so immediate action is critical.

If you don't have an AED, and the individual's heart has stopped, your options are limited. CPR is life sustaining, not life saving, and unless someone receives CPR (and defibrillation) within the first 3 to 5 minutes following cardiac arrest, you may be able to start the heart again but the brain will have suffered serious damage. Also, emergency responders are taught not to put themselves at risk, so the individual must be decontaminated first; this is especially critical at higher levels of containment and dangerous emerging infectious diseases. The best practice is to immediately decontaminate, evacuate, and begin CPR, then wait for responders to arrive with life-saving technologies. The illustration below presents these response scenarios.

There are many emergency situations we can discuss—failures in HVAC systems and safety equipment, along with natural disasters, workplace violence, and terrorism—that require preparedness planning and response strategies. However, I hope you are convinced that it's not enough to only provide plans. Staff must be engaged and must practice plans to ensure that when an emergency occurs, they are prepared and empowered to respond. Their responses could be the difference between life and death, not just for them, but for their coworkers as well as the public.

BIOSAFETY *in the First Person*

We All Have a Bucket by Tim Trevan

> *I first met Tim Trevan in Abu Dhabi, at the meeting he mentions below. Tim is one of the most brilliant people I know. His ability to assimilate concepts and ideas and his understanding of complex systems make him a genius in my opinion. At one time, Tim served as an Iraqi weapons inspector, and he is the author of some very interesting books. Mark my words—it is just a matter of time before Tim makes an impact in the industry of safety.*

Nothing can prepare you for the cave paintings at Font-de-Gaume in southwest France. Unlike the more famous Lascaux site, you can go into the original cave and see the real thing. Only 180 people are allowed in each day, so it's a real privilege.

Your visit starts with a short walk up a narrow, wooded gully to the cave entrance. You enter the cave, perhaps surprised at how big it is, only to realize that this is just where you store your bags and anything else they won't let you take into the caves. Unencumbered, you enter a smaller cave off the one you are in. You walk the narrow path into the deepening darkness, anticipation and excitement growing. Then you get to see the paintings, completed some 19,000 years ago. You see bison, horses, antelope, and mammoth in full color. There are more than 200 "postcards" from the past.

Even as two-dimensional paintings, they are beautiful. But that is only the beginning of how they amaze. The ancient artists painted them in three dimensions. Natural bumps and dips in the rock face are painted to reflect the anatomy of the animals painted. A bump here is the bulging shoulder of a bison. A dip there is the hollow haunch of a horse. And then, la pièce de résistance: if you extinguish all the electric bulbs and look at the images lit by strategically placed candles, with the breeze passing through the cave making the flames flicker, the images come to life. The dancing candlelight animates the scene.

As I was drawn into those extraordinary paintings, my unquestioning arrogance was rocked. I was not a more advanced human than these forebears. The intellect needed to create these images was every bit the equal of mine, some 800 generations of scientific learning and technological advances notwithstanding.

What has this to do with biosafety and biosecurity? I'll get to that. But first we must examine another few threads in my cloth.

In 2014, the CDC was reported to have sent gain-of-function (extremely transmissible) flu virus through the U.S. mail. In the same year, it was discovered that the Proving Ground at Dugway had, over a three-decade period, been sending out anthrax bacilli to 57 laboratories around the world using a faulty sterilization technique that could leave the anthrax viable. How could either of these gross violations of standard practice have happened? These institutions have the world's best experts in biosafety. They are sticklers about compliance. And the United States has some of the world's most extensive and intrusive regulations concerning the handling of dangerous pathogens. It clearly was not a lack of knowledge or awareness.

Likewise, these institutions have engineering infrastructure and safety budgets to make the rest of the world green with envy. If these flagships, these standard bearers, these benchmarks can screw up so badly, how can the rest of us, working under far less ideal conditions, not make such mistakes? And yet many don't.

Clearly, U.S. levels of knowledge, technology, engineering, and budgets are not sufficient or necessary to guarantee biosafety and biosecurity, even if we think so. Just as Font-de-Gaume helped me see that science and technology are not the same as intellect, these incidents at CDC and Dugway clarified for me that biosafety knowledge, technology, engineering, and safety budgets are not sufficient for highly reliable safety performance. To improve safety performance, we must understand why people do what they do.

On February 1, 2017, I heard the British comic Ricky Gervais (a great philosopher in my opinion) explain his belief in science in response to Stephen Colbert's taunt that science is just faith in other scientists.

> "If we take…any holy book and destroy it, in a thousand years' time, that wouldn't come back just as it was. Whereas if we took every science book and every fact and destroyed them all, in a thousand years they'd all be back, because all the same tests would [give] the same results. I don't need faith in science."

Universal laws govern science. Science is about continuous incremental improvement to the body of knowledge. This requires "lightly held" beliefs that yield to contradicting evidence to produce a better model and a deeper understanding. There are also universal laws of human behavior that transcend nurture and culture. We must know these.

In 1990, in a modern classroom overlooking the forest at Fontainebleau that was once King Louis IX's hunting playground, Ingemar Dierickx handed out an assignment to the students in his famed negotiation analysis class.

There is an envelope. You know it contains one $20 bill. You have two days in which to place a single bid for that $20. Your single bid will then be bid one-on-one against the bid of each of the other 89 students in the class. Your total winnings will be added up, and the winning bid and bidder announced in the next class. Thus, if I bid $19.95 and someone else bids $19.90, I win that bidding war and $0.05 (the $20 in the envelope, minus the bid I paid for it). The person who lost the bid to me gains and loses nothing (no $20 in an envelope, but also no bid paid).

This assignment puzzled me. Who would bid anything other than $19.99? Anything more and you would win nothing or lose money. Anything less and you'd lose to the person bidding $19.99. So I sought out Ingemar in his office. He laughed and asked if I was a mathematician or an engineer and was surprised to hear that I had been a diplomat. He hinted that I should think about whether everyone would analyze the situation as I had and, if not, to think about distribution curves.

A lightbulb went on over my head. Not everyone will bid $19.99! I had made an unwarranted assumption that everyone would view the problem the same way. I won't bore you with the details, but six iterations of analysis later, using distribution curves, I ended up bidding just $1.99. I was too aggressive. Had I stopped after three iterations, I would have been close to the winning bid of around $3.49.

Vive la différence! People are different. Psychology and personality matter. To predict behavior of groups, we have to input different perspectives.

Another thread came in 2007, at the First Biosafety and Biosecurity International Conference held in Abu Dhabi, United Arab Emirates. After hearing several distinguished international experts talk about biocontainment, biosafety cabinets, and personal protective equipment, a gentleman from what was then the south of Sudan stood up and said, "This is all very good. But I don't have an advanced lab. I have a bucket, and I'd like to know how I can use my bucket safely." My insight here: we all have a bucket. Regardless of our location, regardless of what pathogens we are handling, regardless of our budgets and our engineering infrastructure, we all have to do what we have to do as safely as we can with what we have got.

Five threads have woven together to shape the way I think about managing biological risk.

- Humility and respect: People nearly 20,000 years ago had as advanced intellect as we do. The same holds true across today's continents; technological prowess is just that. We are no better than they.

- Behavior: Safety performance needs more than just technical knowledge, techniques, engineering, and budgets. The key is behavior.

- Universality: Relativism has no place in safety. There cannot be one sufficient safety for the wealthy/advanced and another for the poor/developing. Equally, across cultures, there are universals in what motivates behavior.

- Diversity: When faced with identical facts, people will analyze them differently based on their knowledge and experience.
- Necessity: We all have a bucket. People everywhere have work they must do, regardless of what they have available to them to do it.

How do we blend these five partially contradictory threads together? Not by dictating the details and specifics, rules and regulations, recipes and checklists, but by promoting a guiding approach, an overarching philosophy, a code that is at once universal and diverse and flexible and adaptive. So, here is my safety creed, my Ten Commandments of Continuous Improvement.

1. Be humble. You cannot improve unless you admit you can be better. You cannot learn everything on your own; you have to depend on learning from others. Nearly everyone, even those we find hard to respect, has something to teach us.

2. Welcome diversity. There is no one true, right way to safety. Stop telling others how they should be safe. Give them the tools to find their own way.

3. Take responsibility. We cannot hide behind lack of leadership, budget, or resources. We can be better with what we already have. We can lead our own change.

4. Learn to lead. A true leader helps others succeed, rather than telling them what to do or how to do it.

5. Know what motivates. To influence others, we need to understand the neuroscience and psychology of cognition (how people become aware of and think about or model their environment in order to predict how their actions will impact their future), decision making (how people use different decision-making methods in different situations), and action (why our actual actions sometimes turn out differently than intended).

6. Know how people learn. We cannot hope to help people change their habits and behaviors unless we know how they learn and the most effective ways of disseminating vital knowledge and best practice.

7. Know your people. To understand what influences our bosses, our peers, and those who report to us, we need to know them as whole people, not as work robots. "It's business, not personal" is nonsense. Everything is personal.

8. Set them up for success. Research shows that, when a person makes an unintentional error in the workplace, the vast majority of that person's peers would have made the same error in the same situation. "Human error" mostly is not. Understand your system and the work environment, and set it up so that it best enables error-free work.

9. Don't think you can plan for everything. You cannot. As soon as humans and machines interact in a changing environment, a complex

adaptive system is created. A feature of these systems, as a result of feedback loops, is that, while certain aspects of the system can be fully understood, the whole cannot. Emergent properties arise. We cannot develop contingencies for every scenario. We can build in some resilience, but even that might not cope with every eventuality. So we must develop our people so they can identify potentially catastrophic developments early and create remedies on the spot.

10. Think. Always think. Kill the "check box and compliance" mentality. We need people to rely less on engineering, standard operating procedures, and checklists and focus more on understanding the fundamentals. We must enable them to identify what can go wrong whenever they act, so that they can devise the safest methods for doing what they must do given what they have. This means that we must trust them to think and act for themselves.

We all have a bucket. Let's learn to use it safely.

chapter 12

Standard Operating Behavior

A standard operating procedure (SOP) is really a standard operating behavior (SOB). Calling an SOP an SOB makes people laugh, so it probably won't become a standard industry term. Regardless, my point is that SOPs are standardized behaviors that are put in place to produce consistent and repeatable safe outcomes among individuals with different levels of education and experience.

During my early years as an infectious disease specialist, I visited many laboratories around the world. At first, my trips were in the United States, and then I expanded my reach to laboratories located in many other developed countries. These countries have laboratories with above average engineering and personal protective equipment (PPE) controls in place. In other words, these countries had the resources to rely mainly on engineering and PPE for risk reduction.

At this point in my career, I was a fan of SOPs and wanted them written for every behavior done in the laboratory. I visited organizations that had thick SOP manuals, and I believed this was a good thing. However, I had become frustrated that even though many organizations had SOPs, very few of them followed the SOPs, and in fact, they had varied operating procedures (VOPs). An SOP is defined as two or more people doing the same thing, the same way, to achieve the same results. A VOP is two or more people doing the same thing, in different ways, and achieving what looks like a similar result, yet when examined closely, both the result and outcome vary. The problem was that most laboratory staff I observed did not behave consistently, even when there was an SOP.

I have mentioned in chapter 5 my experience in a very small laboratory in Honduras. Observing unusually excellent consistency in behavior among the staff, I asked if I could see their SOP manual, thinking they might have discovered something that I could use to address the issue of inconsistent practices among laboratory staff. To my surprise, they had no SOPs other than a few steps written on a sticky note. I was stumped. I asked them, "How do you get consistent behavioral outcomes among different laboratory staff with no written SOPs?" Their answer changed how I would look at and teach SOPs from that point on.

Let's fast-forward to Africa and the Ebola outbreak in 2014. I had just spent a long day working in Monrovia, Liberia, when I received an email from my good friend and mentor Henry Mathews. In his email, he asked if I had seen the SOP that the Centers for Disease Control and Prevention had put out for donning and doffing PPE for health care providers treating Ebola-infected patients. Henry had attached the SOP to his email, and when I read it, I was shocked. All my mentors had come from the CDC, and they were the ones who had taught me the proper ways to don and doff PPE. However, I believed that the SOP I was reading (see Figure 12.1) would increase risk rather than decrease it, and unfortunately, I was right. This SOP was dangerous for many reasons. First, look at how gloves are being removed—this process could contaminate the uncovered finger. This contamination (which I have witnessed during hundreds of glove removal exercises with GloGerm) can inoculate anyone dealing with blood-borne pathogens through tears around the cuticle of the fingers. Second, if you are working with a blood-borne pathogen, why would you remove your eye protection before removing your gown? Third, after removing your contaminated gown, you then touch your face to remove your mask. Meanwhile, handwashing is performed between steps only *"if* hands become contaminated" (my emphasis)—and how would anyone know whether they were contaminated or not? I have stated frequently my opinion that this SOP is dead wrong and irresponsible. To CDC's credit, they are in the process of revising and elaborating on this SOP (https://www.cdc.gov/vhf/ebola/pdf/ppe-poster .pdf; https://www.cdc.gov/vhf/ebola/pdf/ppe-poster.pdf).

In October of 2014, Nina Pham and Amber Vinson, two nurses who followed this SOP while working in Texas with an Ebola-infected patient, became sick. The SOP they were using pushed the boundaries of safety ethics, because they were asked to participate in behaviors that increased their risk rather than decreasing it.

The Honduras laboratory visit and the Ebola outbreak taught me the two most important aspects of SOPs. First, it is not having an SOP that matters; it is how that SOP is transferred from paper to the workforce that counts. Second, organizations must ensure that their SOPs reduce risk; asking people to follow an SOP that increases risk is unethical and dangerous. The troubling part about SOPs is that if they are not managed properly, they can hinder an organization rather than advance it. SOPs are written for two main audiences, regulators and infectious disease pioneers. In most organizations, the regulators are the only ones with the power to address issues found in the SOPs. The infectious disease workers who use the SOPs rarely have the power to revise, modify, and edit them. Most organizations provide for an annual review, but as a behaviorist I can assure you that an SOP can be modified, updated, and improved to increase safety within the first 30 days of its being implemented. Instead of a process that

Figure 12.1 Former CDC protocol for removing PPE (2014)

encourages quick feedback, most organizations will write an SOP, approve it, and then hold their staff accountable for following it as written. This does not change even when the SOP is five years old. Seriously, if an organization hasn't asked its staff to learn better and safer ways of doing things in five years, that organization is not promoting a culture of safety!

My point is that SOPs should not be carved in stone, nor should they be written for regulators. Regulators read SOPs and make sure there are plans in place to address the most obvious risks. Watching regulators read SOPs is interesting. In most cases, they read them, nod their heads, and check boxes as if the plan alone equates to mitigating the risk. We know, however, that a plan alone does not produce a consistent behavioral practice among different individuals, each of whom has different perceptions of risk and different experience and education levels.

So, what's the purpose of an SOP? SOPs ensure consistent behavioral practices among individuals who have various levels of experience, education, and perception of the risks in the laboratory or health care facility. As mentioned, we know that perceptions of risk drive safety attitudes that will then drive safety behaviors. We know that there are many factors that influence perceptions, the main one being experience. The more experience you have with a risk, without being harmed by it, the less risky it begins to seem. When perceptions change, so do behaviors, and different behaviors among different individuals produce different outcomes. This is unacceptable when it comes to addressing risk and safety within a laboratory or health care facility.

The infectious disease pioneers need SOPs to ensure consistent behavioral practices around risks so that consistent safety outcomes around the risk are achieved. It should be recognized that the greatest risk for an organization is not the biological agent itself but the behavior around the agent. The lack of proper behavior around that agent could put the organization, its staff, and others at risk.

So how do we ensure that effective SOPs are written? It is unethical to write an SOP and integrate it into the workforce if it has not been evaluated, validated, and verified. Active SOPs are those that have been published and integrated into the workforce. Inactive SOPs are in development and under review and have not yet been integrated into the workforce. The levels of SOP activation assist in understanding the evaluation, validation, and verification process.

Finally, who should write the SOPs? This is not an easy answer. Scientific SOPs, those used during the process of doing science, whether it is research or diagnostics, should be written by the laboratory director or health care director, who must adequately communicate expectations to infectious disease pioneers. When it comes to writing SOPs for safety, this should also be done by those who are being asked to follow the SOPs. However, safety SOPs should always be reviewed by safety officials. This ensures that both the safety officials and the practitioners are satisfied with the plan. Additionally, a safety official can also ensure consistency with industry standards and compliance with federal or state laws.

SOP EVALUATION

SOP evaluation occurs when the SOP is in the inactive state. It is a three-step process that includes (i) making sure that the SOP reduces risk, (ii) making sure that the SOP is understood by everyone doing the work, and (iii) making sure that the SOP is physically possible. SOP evaluation is not complete until all three steps have been finished.

Does the SOP Reduce Risk?

I have had the privilege of traveling around the world and observing SOPs in action. I have witnessed behaviors that common sense alone would tell you are not part of an effective and ethical SOP. SOPs should reduce risk among those being asked to follow them; they shouldn't increase risk. An SOP that asks health care providers working with a blood-borne infectious agent to remove gloves first and then remove other contaminated PPE lacks basic common sense and judgment.

The illustration above shows the CDC's original SOP for doffing PPE when working around biological agents that are blood borne. You can test it yourself using GloGerm or chocolate sauce, similar to what Sanjay Gupta from CNN did on air (https://www.youtube.com/watch?v=R2ESFx0Fir4). The result is that GloGerm and chocolate sauce are everywhere; this is certainly not an SOP that reduces risk. Make sure that if you write a plan to protect people, the plan does just that.

Is the Plan Written in Language That Is Understood?

After we determine that the SOP reduces risk, we must present it in a form that can be understood. This means that we must review it for language and terminology issues to ensure that the infectious disease pioneers can understand what is expected of them when attempting to learn and follow the SOP.

As mentioned, an SOP is written for two primary audiences, regulators or leaders and workers (infectious disease pioneers). I think the infectious disease pioneers are the most important audience for an SOP, but time and time again, I have found myself in laboratories or health care facilities where the staff can barely speak English and the guidelines that they are asked to follow to protect them are written in English. This makes no sense. SOPs must be written in a language that is understood by the user.

Also, the words in the SOP must be screened for "action verbs" that may be easily read but not necessarily understood. For example, if a cookbook calls for whipping or beating an egg, what does that mean? Do we hit the egg with a whip or our fist? I had no clue prior to writing this book that whipping an egg introduces air into the process and is used for creating a lighter product, usually frostings and soufflés. Beating an egg is more about mixing and is used for structure and "puff" (another example of a word that means something to somebody, but certainly not to me). It's essential to remember that, when we ask people to follow an SOP, there will be greater compliance if they understand the SOP.

Is the Plan Physically Possible?

I often receive a call from an organization with a complaint that staff are not following SOPs. I ask them if the SOP is effective and understood. "Of course, Sean. We would not put SOPs in place that don't work and can't be understood." However, we then discover that the SOPs that are in place are not physically possible—a major SOP failure.

SOPs are words on paper. To demonstrate how easy it is to write an SOP that does not actually work, please try the following SOP.

1. Obtain a sheet of printer paper.
2. With your hands only, fold the piece of paper in half nine times.

That seems simple enough, right? But unless you have the strength of an industrial machine, what is being asked is not possible.

SOPs must be effective, must be understood, and must also be physically possible. I have also witnessed that SOPs for one laboratory or health care facility are adopted for use in another. That is not necessarily a problem as long as you keep in mind that the risks that SOPs are used to mitigate are sometimes very different from one location to another. These differences may not be with the work or the agent but the environment, setup, engineering, and culture, any of which could make that borrowed SOP physically impossible to follow. Finally, consider the individuals being asked to follow the SOP. Physical capabilities within staff can change quickly. If an SOP requires a certain physical trait, you will need measures ensuring that "fit-for-duty" status is retained as an important control to ensure compliance to the SOP.

When the person writing the SOP has reviewed and assured (i) effectiveness, (ii) clear language and terminology, and (iii) physical capability, the SOP passes evaluation and then must then be validated.

SOP VALIDATION

One of my least favorite classes in graduate school was statistics, but it taught me about error in scientific studies. Two types of error are random error and systematic error. I learned that, in general, a larger sample size controls for random error and SOPs control for systematic error. If random error is controlled for, the results can be generalized as the sample size increases. If systematic error is controlled for, the process can be repeated and it will produce the same results no matter who does the process. Isn't this exactly what we want, for SOPs to be generalizable to all people—regardless of experience and education—and for SOPs to produce consistent behavioral results, so that reproducible and predictable safety is the outcome of following an SOP? Of course! SOPs must be internally and externally validated to ensure that they can be repeated, retaining the same results, and be generalized to a diverse workforce.

Internal Validity

At this point, the SOP that has been evaluated remains in an inactive state, meaning that it is not public and is still considered to be under development. Next, the author will need to secure volunteers to assist in the validation process. I usually like to secure 10 volunteers, because I believe that an 80% validity rate is reasonably sufficient. I say this because words on paper alone will usually not produce consistent behavioral outcomes. I would never recommend that the SOP be provided as a replacement for training or mentoring. However, an SOP should be written so that the user can easily understand what is being expected and can complete the task at a rate of 80% or higher. When you have 10 volunteers, the math is easy.

To begin the validation, separate the volunteers into different areas so that they can hear the author's instruction but cannot see each other's behavior. Remember, an SOP is truly an SOB, so we are not asking people to think or feel, we are asking them to behave per instructions. Making sure the volunteers cannot see each other is very important because as the author reads each step of the SOP, one at a time, he or she must ensure that 80% of the volunteers demonstrate *exactly the desired behavior of that step*. If 80% of the volunteers do not achieve the behavior the author desires, the author revises the step(s) and tries again.

After the author has done this with each step of the SOP, and each step has achieved an 80% or greater level of consistency between all volunteers, the SOP has successfully been internally validated.

External Validity

At this point, the author has ensured that the SOP reduces risk and is understood, physically possible, and systematically sound (internal validation). While internal validation looks at process, external validation ensures that the outcomes among different individuals are consistent and that the SOP can be generalized globally (to other humans).

The SOP remains in an inactive state (meaning it is still a draft), and the author is no longer concerned with the behaviors exhibited when the steps of the SOP are read. This time the author will read all the steps and focus on the outcome of the SOP, ensuring that at least 80% of those following the SOP achieve the same outcome. This process is easier said than done. The example described below illustrates why this step may be the hardest of all. Sometimes risks cannot be seen. This is confirmed by the literature, which states that up to 80% of all laboratory-acquired infections were caused by unknown events (Harding AL, Byers KB, Laboratory-associated infections, *in* Fleming DO, Hunt DL [ed], *Biological Safety: Principles and Practices* 4th ed, ASM Press, Washington, DC, 2006). People are not walking around ignoring risks; they are just not seeing them. If an SOP asks someone to remove a glove, well, there are thousands of ways to remove a glove. Each way requires a unique set of behaviors, which in turn produce a unique outcome that may or may not be apparent. External validation ensures that the SOP does exactly what it claims to do in 80% of those who are doing it. When the author has achieved this, the SOP is now ready to become active and integrated into the workforce.

Example of SOP Evaluation and Validation

In 2006, I had been leading the Science and Safety Training Program at Emory University for two years. Each time we facilitated a training program, I used GloGerm to test how well participants removed their gloves without contaminating their hands. GloGerm is a product that is normally invisible but glows under a UV light, allowing individuals to see what they normally cannot see (invisible microbes). Remember, asking people to remove their gloves is a VOP: two or more people do the same thing (remove gloves) a different way and achieve different results. I noticed that around 70% of participants had GloGerm on their hands after they removed gloves, and often on different areas of their hands. This is very concerning, as we know that up to 80% of all laboratory-acquired illnesses come from events that are unknown (Harding and Byers, 2006). I began thinking about developing a process for removing gloves that consistently prevents the contamination of hands.

Over the next 10 years, I developed and modified a process for safely removing gloves and named it the beaking method. Removing gloves via the beaking method results in consistently uncontaminated hands. All infectious disease pioneers can support one another in proper glove removal because the process is easily observable and requires the formation of the "beak." The beaking method is described in detail in chapter 14. This new SOP required evaluation, validation, and verification.

Step 1 is evaluation. We used a room with four wall partitions for five volunteers, who could not see each other because the partitions screened them from each other. Each participant's gloved hands were coated with GloGerm, and the lights in the room could be easily turned on and off, allowing us to observe whether each participant had successfully completed the glove-doffing step by not getting the GloGerm on their bare hands. Another leader and I demonstrated the SOP and asked each participant to follow each step as we did it. The demonstration ensured that they understood the language, that the beaking method was physically possible for all the participants, and that, after the lights were turned out, the beaking method was effective in enabling them to remove gloves without contaminating hands. SOP evaluation was done.

We then moved on to internal validation. Each step was read to the participants. Having the participants in the same room but partitioned from each other allowed us to see them without their being able to see each other. Participants were not influenced by each other's behavior and reacted to the words of the SOP. I would read a step of the SOP and watch what each participant did. If the SOP did not produce consistent behavioral practices in at least four of the five participants, that specific step of the SOP was modified until it produced consistent behavior. This process was long and exhausting; however, by the time it was completed, we had learned a great deal about internally validating SOPs.

The next step was an external validation. We coated the participants' gloves with GloGerm, read each step, and then turned off the lights. All participants successfully removed their gloves without contaminating their hands. This seemed too easy, so I coated their new gloves again, turned off the lights, and this time watched each step, seeing where the GloGerm was and if we had any differences in outcome. To my surprise, we achieved consistency in the process as well as in the overall outcome.

Our SOP had been evaluated and validated, so it became active and ready to integrate into the workforce, both cognitively and behaviorally.

SOP VERIFICATION

SOP verification includes two phases. The first phase is cognitive verification. Individuals are asked to list the steps of the SOP in order and discuss the reasons why the order is the way that it is. This is a very important activity, because it allows the learner not only to demonstrate cognitive mastery of the SOP but also to articulate the reasoning behind each step of the SOP. Remember, humans are not robots; they need to be able to ask questions about the behavior so that they can comprehend and rationalize their participation in the behavior.

Once cognitive verification has occurred, the second phase is behavioral verification. This process aims to connect the brain to the body. I have learned through many years of training that just because someone can state the steps of an SOP in order does not mean that he or she can do it. Behavioral verification is obtained when the individual can physically demonstrate the steps of the SOP in the proper order.

An individual is verified when he or she can independently list the steps of the SOP and then demonstrate those steps successfully. Leadership is responsible for ensuring compliance with SOPs. Both positive and negative accountability for SOP compliance must be delivered to individuals to form habits and culture with a specified environment. Having an active SOP is one thing; ensuring that the SOP is followed and sustained over time is another.

SUMMARY THOUGHTS ON SOPs

The profession of biosafety is stuck in the "In SOPs We Trust" period. When I present the concept of SOP evaluation, validation, and verification, people often shake their heads and tell me that this is an impractical process. However, it's not the process that is impractical, it's having hundreds of SOPs to follow. An SOP is a process designed to minimize the effect of an identified hazard on human health. Providing hundreds of SOPs to an individual prior to working with specific risks is a reverse safety issue.

Reverse safety issues are defined as processes that are put into place to decrease risk but instead increase the overall risk. One example of this is when we wear too much PPE: people can trip, fall, become overheated, and even pass out because of the excess. Having too many SOPs is a reverse safety issue because it makes the assumption that all SOPs are of the same protective value. This could not be further from the truth.

Someone who has 20 years of experience can review a thick SOP manual and easily differentiate the SOPs that matter from the SOPs that do not. However, less experienced individuals with less training cannot make that judgment and may focus on all SOPs equally. Clearly, an SOP for how to use the bathroom would not provide the same protective value as an SOP for doffing contaminated PPE. Organizations must address this reverse safety issue by critically reviewing the SOP manual and then either removing the less critical SOPs or color-coding the SOPs by safety importance.

SOPs have a very clear purpose. If the behavior does not decrease risk to human health and safety, consider calling it a guideline or policy. You can probably cut your SOP manual down by 70%, which ensures that staff who must follow the SOPs can, regardless of experience, distinguish the most important risk management strategies from those that matter the least.

Evaluation, validation, and verification processes should occur only with SOPs where one deviation in the procedure could increase overall risk directly to the individual or other individuals within proximity. Thinking about SOPs this way is very practical.

BIOSAFETY *in the First Person*

The Biosafety Profession: An Unexpected Journey by Joe Kozlovac

Joe Kozlovac is one of the wisest people in biosafety. Starting out at Johns Hopkins and leading biosafety efforts at the United States Department of Agriculture show that he is very accomplished. Joe has been mentored by the best, and yet he is a life-long learner, still considering many of those he serves with to be his mentors. Joe has two phenomenal qualities that I admire. First, he is innovative. As a life-long learner, he seeks strategies that can be borrowed from one profession and applied to the profession of biosafety. Joe introduced me to the concept of appreciative inquiry. Many safety professionals are feared because they

*are always looking for what's wrong; Joe found a strategy used in business that
focuses on what's right. I have seen this strategy work around the world. The
second characteristic about Joe, and my favorite, is a quality that all safety
professionals should have: Joe is pleasantly persistent! Joe never quits, and he
doesn't take a "no" personally. He puts his head down and keeps going, until he
pushes safety within his organization up two more steps. My email box fills up
fast these days, but when I see Joe's name, I stop and read. Whether it is an
observation or an article he is forwarding, I continue to learn much from him.*

Just like Bilbo Baggins, I never expected when I graduated from the University of Pittsburgh at Johnstown with a biology degree in the mid-1980s that I was about to set out on an unexpected journey that was life altering. With my new, shiny diploma, I was hoping to get a job perhaps monitoring acid mine drainage water treatment plants. However, I ended up relocating to the Washington, DC, metropolitan area when I landed my first position in a large-scale biosafety level 3 (BSL3) virus production laboratory that provided viral antigen for HIV and HTLV-1 *in vitro* diagnostic test kits. When I told my parents about the job, they were not very happy, concerned that I would become infected performing such work. My father, in an attempt to dissuade me, told me if I took that job to not bother coming home (he later recanted and informed me that I had made the correct decision because it led to my career in biosafety).

The job paid $16,700 a year. I did not know what a BSL3 lab or a biosafety cabinet was and certainly had no experience working in either. I learned from more experienced lab colleagues, including several former U.S. Army Medical Research Institute of Infectious Diseases technicians, and rose through the ranks to eventually manage these labs. It was in this high-containment environment where my interest in biosafety first began to stir along with other ancillary duties related to my production work, such as emergency response and training video production. When a biosafety position was advertised at the University of Maryland at Baltimore, I was pleasantly persistent in expressing my interest in the position until I was hired as the deputy biosafety officer. It was at this point that my true education in the field of biosafety began, starting with the first week on the job, during which I attended the Control of Biohazards course taught at Johns Hopkins University School of Hygiene and Public Health by Byron Tepper and Richard Gilpin. Byron and Richard later became mentors, friends, and my supervisors when I migrated to Johns Hopkins a few years later as associate biosafety officer. During my years at Johns Hopkins, I developed as a biosafety professional, and it was this foundation that prepared me for the challenges of being the institutional biosafety officer at the National Cancer Institute in Frederick, Maryland.

The timing of my entrance into the field of biosafety was fortuitous because so many of the giants in the field—Barbeito, Barkley, Tepper, Richmond, Tulis,

Spahn, Songer, Vesley, and Fleming, to name only a few—were still active and very available to a young biosafety professional with many questions. If there were such a thing as biosafety "baseball cards," I would have been quite a collector. Following the guidance of one of my mentors, Steve Pijar, who stressed networking in the biosafety community, I began creating my own network, which included many of these thought leaders in the biosafety profession. I am very grateful to have had the extreme good fortune of being mentored by them, listening to their stories, and having them as my professional role models over the past 30 years. As a biosafety professional, you need to establish a good professional network to tap for information or even just to bounce ideas off of because, believe it or not, you will not always have the answer and more than likely will not find the answer in a Google search or on a listserv.

After spending 30 years in the biosafety profession, here's what I have learned from my own experience as well as by observing those outstanding biosafety professionals whom I have tried to emulate. First and foremost, over the course of your career you can expect critical review from scientists, leaders in your institute, and even other safety professionals. You should never take this personally. You are at your institution to serve the life sciences and should always remain professional but also know where your line in the sand is drawn. A biosafety professional also has to be prepared to work, even thrive, under pressure from conditions such as incident/accident response, regulatory visits, and very tight deadlines. I can assure you that those moments will come, and if you do not believe you can work well under pressure, then the biosafety profession may not be a good fit. Also, do not give up; always be pleasantly persistent in making your case for your biosafety program. In many cases an institution may not be ready to make the changes you are proposing, and there may be valid reasons for that organizational resistance at that particular time. I have found that if you are persistent and consistent in your message (and, of course, if it is actually a good idea), you can champion change even in the most resistant of organizations. It will not take anyone by surprise that one of my favorite quotes is by Calvin Coolidge, which starts, "Nothing in this world can take the place of persistence…."

Technical or professional credibility is critical to the biosafety profession. Without exception, leaders in the biosafety field whom I have observed are technically proficient in the core competencies as defined by the certification exam, e.g., risk assessment, disinfection and sterilization, biosafety program management, etc. But they did not stop there; they further distinguished themselves by developing expertise in specific areas of biosafety, e.g., high-level containment, agricultural biocontainment, field applications, and laboratory design. In most cases they have published in these areas as well as taught others on the subject. A focus on continual learning is a key component to becoming a successful biosafety professional. Any success I have had in this field is primarily due to a persistent focus on *becoming* rather than *being*, which I believe is a strategic principle shared by many leaders in general but is very pronounced in those considered thought leaders in the biosafety field.

Communication skills in all forms—writing, public speaking, interpersonal communication, and conversation—are critical for the biosafety professional and also the essence of good leadership. Biosafety professionals at our most effective can serve as institutional glue for those we serve because we communicate to all levels within the organization, from housekeeping staff to the corporate, university, or government executives. We can reinforce institutional values related to the biosafety program. Without good communication skills, you are not going to have influence or impact within your organization, and you will not inspire anyone to embrace the tenets of biosafety. Communication, especially public speaking, was a major hurdle for me to overcome. I had extreme difficulty—in fact, crippling fear—getting up in front of a small group to introduce a training video, but I knew that if I wanted to go forward as a biosafety professional I would have to overcome this. Therefore, I threw myself into every type of public speaking opportunity, and I bombed a lot along the way. I eventually became comfortable with speaking to an audience of just about any size, even though I am still a little nervous right before I begin any lecture.

Setting a good example, or practicing what you preach, is another powerful means of communication. When I became a grandfather eight years ago, a poem by Rev. Claude Wisdom White, Sr., came to my attention titled, "A Little Fellow Follows Me," which eloquently states the case for making a good example. I have observed firsthand many times the positive impact that leaders can have when they lead by example. In the 1990s, when I was the associate biosafety officer at Johns Hopkins Institute, we were having some difficulty getting folks to attend their Occupational Safety and Health Administration blood-borne pathogens annual retraining, especially physicians. There were regularly scheduled training sessions on this topic, and during one of these sessions, an individual walked in and quietly took a seat. Everyone in the classroom immediately recognized Michael Johns, dean of Johns Hopkins School of Medicine and vice president of medical faculty, who at the time was also advising the Clinton Administration on health care reform. Dr. Johns sat through the presentation and signed the attendance sheet, his attendance sending a powerful message to me and the staff in attendance relative to the importance of compliance with safety regulations. I kept a copy of that signed attendance sheet up on the wall in my office for years as a reminder of leadership by example.

Developing and mentoring others is another hallmark of good leadership that fortunately is quite prevalent in the biosafety community. I have been blessed to have some great supervisors and professional mentors who believed in me and, more importantly, gave me opportunities that pushed me beyond my comfort zone to develop new skills and abilities. It has often been said by individuals much more knowledgeable than I on the topic of leadership that the best leaders do not have followers but create other leaders. In my opinion, Robert Hawley is a top practitioner and thought leader in the field who has been a true mentor, developing others in the biosafety profession as well as outside the profession. I would not be where I am today without the opportunities and wise counsel Dr. Hawley provided to me over the years, encouraging and providing the proverbial kick in the backside when I was hesitant to take on a new challenge.

It is imperative, especially with the increasing need for biosafety staff at life science institutions, that experienced biosafety professionals pass on what we have learned to those new to the field and provide them to the extent possible a safe environment in which to make errors and practice leadership. I also believe that it will become an increasingly frequent and important role for biosafety professionals to be able to actively coach and mentor the senior leaders within their organizations to become biosafety champions and ensure that they communicate the appropriate message, model the behaviors they wish to see in their organization's employees, and determine what metrics should be used to determine biosafety program effectiveness.

This can be a challenge. It takes time to develop the needed level of trust in the biosafety officer among senior leadership. One of the means to achieve this is to be consistent and pleasantly persistent in your messaging, performance, and interactions with leadership. As my friend Jim Welch has often stated, "biosafety professionals lead from the second chair," which means that we are not typically the corporate or institutional executives creating the vision and direction of our institutions, but we do play a critical supporting role to any organization's mission involving work with biohazards. Therefore, for biosafety professionals to be maximally effective, we must develop "soft" management and leadership skills to create key relationships with leadership and the workforce so we can become the institutional glue that fosters a commitment to safe science and an atmosphere of trust within the organization. Unfortunately, training on those soft management and leadership skills has until recently been overlooked in most biosafety professionals' training, which typically focuses heavily on the technical aspects of biosafety and biorisk management.

I think the last and maybe the most important ingredient essential for becoming a successful biosafety professional is passion for the profession. This is not only passion for the practice of biosafety but also being passionate about recognizing biosafety as a distinct profession. The high demand for biosafety professionals may bring individuals into the profession who may not be as imbued with the desire to see biosafety recognized as a distinct and separate discipline as those of us who went through long apprenticeships with seasoned veterans. Another growing concern related to rapid expansion in the biosafety profession is the possibility of an individual without the appropriate training and experience assuming a biosafety officer role prematurely, resulting in harm. This is one of the reasons why formalized training and mentoring programs, such as the National Biosafety and Biocontainment Training Program, are needed, as well as an emphasis on the importance of maintaining high standards for professional biosafety credentials, such as the certified and registered biosafety professional program through American Biological Safety Association International. I believe this will continue to remain a challenge that we as biosafety professionals will need to address.

I would not trade the last 30 years in the biosafety profession for a mountain of gold. I think the role of biosafety officer is one of the most challenging, rewarding, and often exciting positions to have. My colleagues in the profession are family, and it is a good family to belong to and to continue to nurture and grow within. My professional journey, although unexpected, has been and continues to be one of the most rewarding experiences of my life.

The views expressed in this contribution are solely those of the author and do not reflect the views or official position of the U.S. Government.

chapter 13

Effective Training Strategies

It is difficult to train people who think they already know everything. This is a human risk factor and an enormous challenge to safety. How can you prepare someone for jumping into the deep end of a swimming pool when that person believes that he or she knows how to swim but, in reality, does not? Only after jumping into the deep end does this person find that he or she is drowning. Sometimes, it is too late.

Life teaches us real-time lessons, but the price of these lessons can vary. Learning vicariously through the experience of others and experiencing a near miss are lessons that result in little or no personal harm. Unfortunately, the price of lessons can go higher. Incidents turn to accidents, and one can experience a non-life-threatening accident, a life-threatening accident, or an accident that leads to death. Life's lessons keep occurring until the day we die. To learn them, we must assume that we don't already know them, which is not a common attitude.

TRAINING, ROUND 1

For many years I have started my training programs with an exercise that demonstrates the transition from unconscious incompetence—you don't know what you don't know—to conscious incompetence—you know that you don't know. This exercise involves completely putting together the seven pieces of a puzzle that is available from the Elizabeth R. Griffin Research Foundation (see chapter 3).

I ask participants to estimate their likelihood of completing this puzzle in 1 minute or less. I have probably done this exercise for several hundred people around the world, men and women of all ages, some of whom speak English and some who do not. The same results occur more than 99% of the time. Here is how it goes.

I begin by asking for two volunteers, one man and one woman. When both come to the front of the classroom, I show them the puzzle pieces and ask them to tell me, using a scale of 0 to 100%, the likelihood that they can complete this puzzle in 1 minute or less. There has been a strong trend that men are much more confident than women; most men will say 50% and

Figure 13.1 Elizabeth R. Griffin Research Foundation puzzle

most women will say 30%, but of course there are outliers. Whoever offers the highest number is the volunteer I use. I ask all training participants to come close and stand around the lucky volunteer, who is now sitting down with the puzzle pieces in front of him or her.

One time that I did this, my volunteers were Jan, who said said 70%, and Greg, who said 50%. Jan sat down, and when she was ready, I asked her to put the puzzle together. After 1 minute, her attempt failed. I said, "Now that you have tried this once, what would happen if I gave you another minute to start over and try again. What is the likelihood that you would complete this puzzle in 1 minute or less?" Jan said 50%. I know that a 20% reduction is not enough, but I gave her another minute. Jan tried a second time and failed. Again, I asked her to rate the likelihood that she could complete this puzzle in 1 minute or less. This time, she said 0. BINGO! She had a transition from 70% to 0, moving from unconscious incompetence—I know I know (but really don't know)—to conscious incompetence—I know I don't know.

There are some learners who can quickly differentiate what they can and cannot do, and others who need substantial periods of time before realizing what their personal limitations truly are. It is not uncommon for someone to try the puzzle twice, but some have tried six or seven times before admitting that they cannot put the puzzle together in 1 minute or less. I call these stubborn learners, which I consider a compliment. This is where I will pause the story to make some important points about learning.

NOT EVERYTHING IS TRAINING

My experience has shown that learning occurs through three formal processes. If we provide information and facts, we are offering an *awareness session*. This kind of program increases knowledge about a topic. If we provide an opportunity to put that knowledge into practice, building a skill that can be observed and evaluated by a third party, we are offering a *training session*. Training builds skills and abilities of those in the program. If we provide challenges to participants that ask them to blend their experiences and abilities, a blending of both cognitive and behavioral lessons, we are offering an *education session*. Education increases the resourcefulness and application of both ability and cognition by applying lessons to real-life situations. So learning can be awareness, training, or education. Not everything we do is training, and the use of the word "training" loosely is damaging and potentially leads to a greater resistance in learning.

NOT ALL LEARNERS ARE EQUAL

Although I would like to believe that all people who attend a training program want to learn, this isn't true. Some show up to learn, others show up to evaluate what the trainer is teaching, and some attend to teach the trainer! I have found there are three types of learners: novices, practitioners, and experts.

Novices show up to learn. They are eager, they have no experiences that lead them to challenge or teach the trainer a different way than what is being taught ("contamination"), and they learn very quickly.

Practitioners show up to teach. They have experienced early success in their careers and challenge anyone who tells them to do anything different than what has led to their success. It is usually a "my way or the highway" mentality. They argue, debate, and question what the trainer is teaching. Practitioners are the hardest group to train.

Experts are the hardest group to change. They attend a training session with their arms folded, nodding here and there with the confidence they have earned and deserve. Experts have been successful for many years and during that time have realized that there are many ways to achieve success. They have accepted that what works for one may not work for another and have no concerns with challenges to their own personal processes. Experts can become difficult when being forced to change, but a proper training avoids that. Learning is one thing and changing is another, and they should not be confused or combined.

MOTIVATING LEARNERS

I have distinguished the different types of learning and learners, but even among the differences are commonalities. No matter what kind of learning experience you are offering to a learner, because we are teaching humans, we must "marinate" to improve the learning process. To marinate is to make learners hungry to learn and motivate them to receive the lessons we are attempting to provide them. This is where understanding human needs—specifically, belongingness and self-esteem—helps. Challenging basic needs like self-esteem and belongingness produces the desire to behave. Once you have a desire to behave, you can begin to direct specific behaviors to achieve the results you are looking for specific to the overall learning process.

Have you noticed your behavior when you are given a test that is supposed to evaluate your knowledge and ability? You become serious. You focus on the test and rarely evaluate or judge the questions or tasks that have been asked of you. You trust that the questions being asked are fair, and if you fail, the initial failure gets your attention. It is just possible you don't know as much as you think you should. If you are given the same question again, and miss it again, this creates a hunger to find out the answer. How could you miss the same question twice?

How would you feel if you missed work for an hour to attend a training session, and because you did not pass the program you had to attend the session again? Asking a simple question like this before a training program can increase the desire to behave. As a trainer, if you direct people to behaviors that increase their likelihood of succeeding, they are more likely to comply and behave. These concepts are important and the reason I almost always ensure that testing and certificates of completion are distributed at any learning program I am facilitating. They are drivers of motivation for the learners.

TRAINING, ROUND 2

Now we will return to the story of the training exercise, where Jan has just realized that she didn't know as much as she thought she did. Her experience was equal to failing a pretest. Although she may have come to the training to evaluate, judge, or replace the trainer, Jan has learned that she has something to learn and is now more eager to listen, which facilitates learning. I told Jan that I am going to teach her the standard operating procedure (SOP) for putting this puzzle together. I told her that it must be done exactly as I will teach it. Jan nodded.

I began putting the pieces into the puzzle frame. At the fourth piece, I stopped and said, "This is the hardest piece of the puzzle, so I will show you how to do it and then I want you to do it." I showed Jan how to do it and asked her to demonstrate the behavior. This really isn't the hardest piece, but it is the middle piece of seven pieces, and to keep learners engaged, they must be involved in the learning process. After Jan put the piece in, I finished assembling the puzzle. Then I dumped the puzzle pieces out again and announced, "the second round of training."

The title of this chapter is "Effective Training Strategies." Therefore, we are discussing the best way to train or teach behavior and skills. What is the difference between teaching and mentoring? There really should not be any difference, but in general, teaching is a one-way learning process, whereas mentoring is a two-way process in which the learner is actively involved in the learning process. When one teaches, failure is a measurement of success. When one mentors, failure is part of the process. You cannot learn unless you fail, and in failing, you tend to learn even more.

In this example, Jan has learned two key points. First, the puzzle can be assembled; sometimes participants wonder if there is some kind of trick to it. Second, Jan has learned the SOP for the first time. Can you guess how many steps someone remembers of a seven-step SOP the first time they learn it? Let's see.

I started by laying all the pieces in front of Jan. I told her, "For this round of training, point to the piece that goes in first." Jan pointed to the first piece. I put the first piece in. She then pointed to the second piece, which I inserted. Then Jan stopped and looked at all the pieces; she then pointed (as often happens) to the piece I had her put in, which is the fourth piece, not the third piece.

During emergency response situations, I have been asked to assess who can learn an SOP quickly and who needs more time. I find that roughly 70% of individuals will not be able to tell you the third step of an SOP after learning it one time. The others who can, the 30%, tend to demonstrate increased abilities in both learning SOPs and attending to the details needed to follow them. This is also important because we should recognize that SOPs are not mastered by just doing them once. Learners need repetition for learning to take place, no matter whether they are in awareness, training, or education sessions.

At this point, Jan had just identified the wrong piece. I told her that is incorrect and asked her to guess again. Now, Jan begins to struggle.

The most important attribute of a trainer, someone who is teaching someone else a behavior, skill, or ability, is that they allow the learner to become comfortable with struggle. I cannot stress this enough. If the trainer jumps in and rescues the learner and provides the answer, the next time the learner struggles, they will look outside themselves for the answer rather than figuring it out for themselves. As in Gandhi's maxim, trainers do not provide the fish that feeds someone for a day, they provide the skills needed to fish that will provide food for the rest of their lives. Trainers must be comfortable with allowing learners to struggle because it is within the struggle that the greatest leaps of learning occur. Please note that struggle can lead to frustration, which then leads to someone giving up. Never let someone struggle to the point where he or she quits; that is too much struggle. So, observe the struggle, and when they get frustrated, provide support. When they lose hope and are about to quit, give them a clue. Be there, but remember that your job is not to do something for them but teach them to do it for themselves.

Someone from the audience pointed to the piece that Jan needed. I reminded the audience that when Jan does this SOP in the future, they may not be there to help her and so to please let her struggle until she figures it out herself. Jan pointed to the correct piece, and we continued. We got to the fourth piece, and again I asked to put in the hardest piece, which Jan did with success and then smiled. I asked her to complete the rest of the puzzle, which she did successfully. I then dumped all the pieces out again and said we would move to the third round of training.

TRAINING, ROUND 3

I asked Jan to point to the first piece of the puzzle and then point to where it goes. Can you guess what she did?

I discussed making someone hungry to learn. At this point, I had taught the learner the same SOP twice. What do you think she is feeling? "I can do this. Get out of my way and let me show you." What a great feeling this is. There are many ways to assemble this puzzle but nobody really talks about that. All they want to do is show you they can do it the way you have taught them to do it. How do you know that she is eager to learn?

Rather than point at the piece, Jan grabbed it. I immediately asked her to put the piece down. I requested that Jan point to the piece and then point to where it goes. She did this successfully for the first three pieces. Beginning with the fourth piece, I asked her to complete the rest of the puzzle.

When teaching an SOP or behavior, you don't need a PowerPoint or words on paper. Have someone go through three rounds of training. In the first round, the person watches you do it. In the second round, they identify the order of the pieces as they watch you assemble the puzzle. In the third round, they identify the piece and point to where it should go. At this point, you really want to increase their confidence, so what is the fourth round of training?

TRAINING, ROUND 4

I reminded Jan that this is an SOP, and it is important to follow all steps in the proper order. I told her that the fourth round of training is like playing a game of ping pong. I will put the first piece in, and she will put the next piece in. We will rotate putting pieces in until we complete the puzzle.

However, I intentionally placed the sixth piece in when I should have placed the fifth piece in. Jan said nothing, grabbed the fifth piece, and placed it in the puzzle.

Good science is safe science. Great science is replicable. This means that if I do the same science in one laboratory in the United States and one in Pakistan, Mexico, France, China, Thailand, or Canada, it produces the same outcome. Well, safety is the same. Great safety is replicable, and the SOP is the recipe for replication. The fourth round of training builds confidence but also teaches the learner to speak up when someone breaks the SOP. Never assume that teaching the SOP will provide the skills and initiative needed to confront another person who may not be following the SOP. Always teach learners that safety is not an individual action but a family one. We are only as safe as the riskiest behavior around us.

"Hold on," I told Jan. "I didn't follow the SOP, and you didn't say anything. Let's do it again." We repeated the process, and again I took the wrong piece. Immediately, Jan identified this and corrected my behavior. We continued and completed the puzzle. Jan smiled. Next, she faced the final round of training.

TRAINING, ROUND 5

One could argue that Jan is now trained; that is, that Jan has successfully transitioned from unconscious incompetence, thinking she knows what she doesn't know, to conscious competence, knowing that she knows. However, because Jan is human, she has not made the final transition. Whenever humans learn a new behavior, they become very egocentric. They believe their success at completing a behavioral request has everything to do with their abilities and fail to recognize outside influencers.

As Jan placed the final piece in the puzzle, I asked those watching the demonstration to give Jan a round of applause. I asked Jan, "On a scale of 0 to 100%, what is the likelihood that you would be able to put this puzzle together, using the SOP you have just learned, in 1 minute or less?" Jan quickly answered 100%.

This is Jan's way of saying "I can do it!" There is no doubt that she can. After all, she has watched it three times and then assisted doing it two more times. Most humans at this point can put together a seven-piece puzzle in less than 1 minute using an SOP they have watched five times. So how do we obtain a level of conscious competence that recognizes much more than skill and ability?

As soon as Jan said 100%, I countered by saying, "Close your eyes." After she closed her eyes, I took a piece of the puzzle and then asked her to begin. She did exceptionally well, as most do, putting all the pieces together even with her eyes closed in 45 seconds. She spent the remainder of the time feeling around for the final piece that was in my hands, then opened her eyes. I announced that Jan was unsuccessful and asked her to close her eyes again. I then asked her to tell me how many pieces were in the puzzle.

Jan has just failed because she could not use her eyes and didn't have all the pieces. However, now I have asked her to tell me how many pieces are in the puzzle. Why am I doing this? When you verify if someone knows an SOP, there are two levels of verification (as discussed in chapter 12). The first is cognitive (the ability to list the steps), and the second is behavioral (the ability to do the steps). After watching Jan do the SOP with her eyes closed and search for the final piece, I have verified her behavioral ability. If she answers the right number of pieces in the puzzle, where she must think about what she would do, she is now cognitively confirmed as well. What does she say?

With her eyes closed, Jan thought about how many pieces are in the puzzle and said, "Seven!" People clapped, and Jan opened her eyes. I then asked Jan, "On a scale of 0 to 100%, what is the likelihood that you will be able to put this puzzle together in 1 minute or less following the SOP?" The golden moment of truth has just arrived. Will Jan demonstrate conscious competence or not?

Jan answered, "Will I have all the pieces and my eyes open?" "Yes, of course," I responded. She said, "100%," and proceeded to successfully assemble the puzzle. Success! The moment Jan realized that success relies on something greater than her own skills and abilities, she transitioned into conscious competence. She recognized that she is capable if she has the resources and right work environment to do what she is being asked to do.

But that is not where the exercise ended. Remember that I had two volunteers. Because Greg had rated the likelihood of his success lower than Jan rated hers, I had asked Jan to stand up and Greg to take a seat in front of the puzzle. Now I said to Greg, "Please assemble the puzzle using the SOP." Greg assembled the puzzle, a little sloppily but recognizing the way it should look and successfully following the SOP. Greg smiled and those around him applauded his success.

The final phase of this exercise demonstrated that humans learn vicariously, by watching other humans. Then why waste a moment training with a written SOP? Can't they learn just by being asked to do it? This final training step demonstrated that when we train, it is best to demonstrate the behaviors we are looking for and then ask those being trained to demonstrate them back to us—repeatedly. By fine-tuning or shaping their replications of behavior, we begin to see mastery, like the transition from being a cook to a chef. Remember, a recipe card alone will never make a cook a chef. An SOP alone will never make someone working with an infectious disease safe. We must train with those learning, to ensure that the skills and abilities needed are mastered before we allow them to work around risks.

CONCLUSION

As mentioned, I have facilitated this exercise hundreds of times, around the world, with men and women of all ages, and yet it elicits a predictable and similar response by all. This exercise demonstrates five critical points about learning.

First, not all information sessions should be called training. Learners have much to learn. They must have awareness about what they are learning. If what they learn requires a skill, they must learn the behaviors needed to do what is needed, when it is needed. Only when they have experience can we begin to challenge them to apply their skills and experiences to solve problems and develop innovative solutions that can prevent accidents, in addition to responding to them. Separating awareness, training, and education so that a trainer can focus on the specific goals of a learning session is critical when implementing a plan for learning.

Second, all learning sessions should plan to include the three levels of learners and to understand them as well. A novice will usually be the easiest to train, because there is no existing knowledge to compete with and new skills are absorbed with little or no friction. Trainers should expect challenges when training practitioners, because practitioners are usually at a stage in their career where failure in work could lead to failure in their profession. Trusting a new process or behavior will take time, so at a minimum the trainer should expect resistance. Experts offer little resistance when attending learning sessions. This is usually associated with their overall levels of engagement. So trainers should engage them, utilize their knowledge and experience as examples, and ask them to serve as an example for novices and practitioners during the training session. Remember, practitioners are the hardest to train; experts are the hardest to change.

Third, humans live in a state of unconscious incompetence and sometimes need to be motivated to learn. Trainers must make learners "hungry," or motivated. Use psychology and market the benefits of successfully completing the training session. This means you must find something that your learners value and, using the basic human needs of belongingness and self-esteem, increase motivation among them. There are countless examples of students performing to win approval from their instructors or their classmates. Trainers can use these basic human needs to motivate learners to succeed in training sessions.

Fourth, when humans learn a skill and ability, they believe it is their skill and ability that determine success. This is a human risk factor, and all training sessions for learning a behavior should ensure that learners become aware of the multiple factors that lead to success. Skills and abilities are one thing, but without proper resources or consideration of other environmental factors, success will be limited even with the best skills and abilities. To support this point, I love to use the example of driving. You could be the best driver, but if the car's brakes don't work, it would be a challenge to be safe. Also, even for the best driver in the best car, an ice storm could challenge safe driving. Finally, if the best driver in the best car is on the road with drunk drivers, the safety of the best driver will be impacted. Like the road, our safety when working around infectious diseases has much to do with our skills and abilities, but it also requires us to pay attention to the resources and the environment in which we are working.

Last, when teaching learners behavior (or facilitating a training), stop reading and start doing. It drives me crazy that we treat humans as though they are computers. Humans learn by doing, and even if those humans state that they prefer to learn by reading, we must consider the overall outcome we are expecting from the learner. If we are teaching a learner how to spell, reading works. However, if we are teaching a learner how to swim, reading about swimming may comfort the learner by being informative, but it's useless when the learner jumps in the deep end. Behave first, and once you are comfortable and have a foundation of the expected behavior, then study the information needed to master the behavior. Humans learn vicariously, by watching, so demonstrate and ask them to replicate. This is an exceptional way of training someone on an SOP.

A live video of this training demonstration can be viewed at https://www.linkedin.com/feed/update/urn:li:activity:6410575115867738112.

 BIOSAFETY *in the First Person*

She Was One of Us! by Karen Byers

I once told Karen Byers that writing a book was on my bucket list. It is Karen who called me one day to encourage me to write a book proposal and submit it to ASM Press for consideration. I thank Karen for her encouragement and support, because I am not as confident as I appear and she certainly deserves credit if you find any value in this book. Karen is an extremely brilliant, witty, and science-driven biosafety professional. She does so much for the profession of biosafety and is a specialist when it comes to laboratory-associated infections. She is an incredible resource, and I am very grateful for her contribution to this book and for what she has given to the profession of biosafety.

I was invited to provide a biosafety workshop to a group of public health and clinical laboratory staff. The session was going extremely well, and I was very gratified by the level of interaction. I presented scenarios that led to laboratory-acquired infections that had actually occurred in clinical laboratories in the United States and Canada. Participants responded when I asked for recommendations on prevention of similar incidents, and the discussions were lively and on point. I went through case studies from published literature, such as *Escherichia coli* O157 accidentally replacing the Gram-negative control for antibiotic sensitivity and the restreaking of a *Brucella* isolate out on the bench that resulted in an outbreak in a large microbiology laboratory. By the time we did a case study about a student who infected several coworkers by contaminating the handwashing sink with *Shigella*, everyone in the room had a story to share about how they had dealt with a student who was skeptical about the basis for the safety practices.

Confident that I was doing a great job and connecting with the workshop participants, I got to some fatal *Neisseria meningitidis* case studies and flashed details from several *MMWR* articles on the screen. The participants' level of engagement suddenly crashed by any measure that you can use during a live training. No one was nodding, most were looking down, and no one was making eye contact. Was a whole morning of case studies just too much? Maybe I had put in too many details, or coaxed out too much discussion? Maybe a whole morning of me was just too much.

Clueless, I soldiered on. I discussed the recommendation in the Sejvar article [Sejvar JJ, et al, Assessing the risk of laboratory-acquired meningococcal disease, *J Clin Microbiol* **43**:4811–4814, 2005] that samples from sterile sites, such as cerebrospinal fluid, be worked up in a biosafety cabinet. In previous sessions, some individuals spoke up at this point to say that they did not have access to a biosafety cabinet; then we talked about how to justify a request, how to politely repeat a request at every budget submission, and what interim safety measures were in place for sterile-site samples when they did not have a biosafety cabinet to do diagnostic procedures with cerebrospinal fluid. I also heard concerns about the off-hour and weekend staff who did not routinely do microbiology but were lab techs who would process "stat" samples in the microbiology lab when required. In all those earlier sessions, we brainstormed about the best methods of communication in their workplace, both at lab meetings and with signs. But, in this workshop group, there was dead silence.

I talked about those responses from previous sessions and then asked if there were any questions or if anyone had additional experiences to share; no one did. The organizer thanked me, then announced the lunch break. Participants clapped and filed out.

I had no idea what I had done wrong. I went up to the organizer to ask how she thought the session went. She seemed agitated and blurted out, "I knew her! Most of us knew her. I had no idea that her case was in the *MMWR* or that you would talk about her. I knew her, and she was one of us."

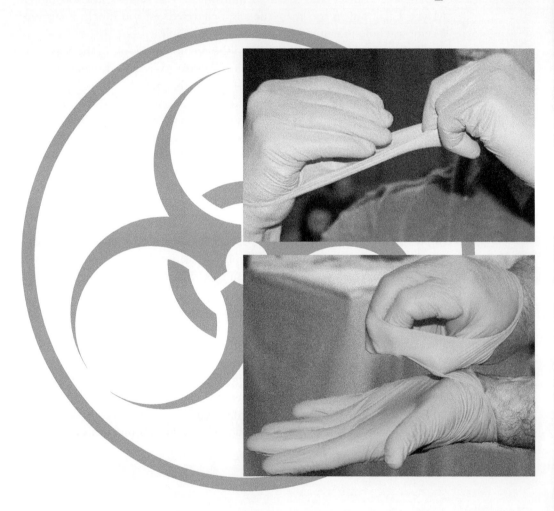

chapter 14

The Beaking Method

It is unethical and wrong to give people personal protective equipment (PPE) and allow them to interact with biological agents while providing a standard operating procedure (SOP) that fails to keep them safe while removing the PPE. During the Ebola outbreak in 2014, the Centers for Disease Control and Prevention made this mistake. Many people continue to make this mistake around the world daily.

We know that at least 70% of all laboratory- and health care-acquired infections come from unknown events, i.e., they cannot be traced back to a specific event. I am convinced, after 15 years of traveling around the world and watching nurses, doctors, and laboratory staff remove their gloves, that it is how they are removing their PPE that leads to many of these infections. To prove my opinion, simply coat your gloves with GloGerm, or other goo of your choice, and follow the beaking method for glove removal. Do it 10 times. Then do any other research-supported glove removal process 10 times. See which method is better, that is, safer when working with biological agents. That is my research. Please consider doing the research yourself. The beaking method is the only one I know of that consistently produces the same result among different people and professions in the world.

The beaking method was developed while I was at Emory University. Lee Alderman, Henry Mathews, and I turned out the lights, coated our gloves with Glo-Germ, and removed thousands of gloves until we developed a process that did not leave one speck of glow on our hands. We then asked our students to demonstrate this behavior, and they repeated our results. To this day, you cannot pass the classes I teach without demonstrating proper and effective glove removal by following the beaking method.

I mention this method throughout this book, and it is depicted in detail below. Any SOP should always be tested to ensure that it reduces risk. So get a pair of gloves and some chocolate sauce and follow the SOP below to remove your gloves without contaminating your hands. (Note that maple syrup or barbecue sauce also work well—any sticky spots represent contamination!—as would powdered chalk, bright-colored finger paint, liquid fabric whitener, and so on.)

Step 1. Form an "L" with the left hand. (This method requires some strength in the hand being "beaked," so persons who are left-handed may find it easier to start with the right hand, or a "J." Just reverse the hands left/right for the steps below.)

Step 2. Place your right hand in the "give me five" position, extended with the palm up. Use your left index finger and thumb to pinch up the cuff of the glove at the inside of the right wrist, making sure not to touch your skin with the gloved fingers.

Step 3. Hook up the glove cuff from the outside with your middle finger, and only your middle finger. Once the glove edge is elevated, hold it between the middle finger and the thumb.

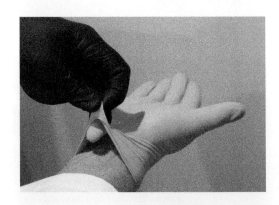

Step 4. Keep the right hand in the "give me five" palm-up position to avoid contaminating the inside of the glove.

Step 5. "Beak" the right hand, pulling the glove inside out to hook over your thumb and fingertips only. Pull only this far; too far, and you will lose control of your gloves.

This should feel very tight. That tightness will always serve as a cue that the clean side of the glove is now covering the outside of your fingertips.

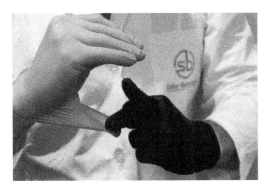

Step 6. Because the inside of the glove is clean, don't grab the other glove where it is dirty. Instead, use the beaked, covered fingertips to pinch up the left-hand glove at the edge of the cuff on the inside of the wrist.

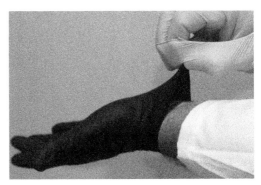

Step 7. Peel off the glove on the left hand. Keep the fingers on the right hand (the beaked one) pointing forward as you do this. This prevents you from contaminating your ungloved hand as the glove turns inside out when removed.

Step 8: Drop the left glove in the trash.

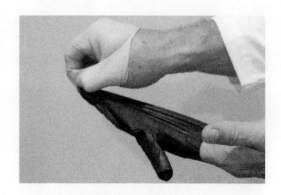

Step 9: Place the right hand in the "give me five" position, stretched out flat and palm up, with the clean inside revealed.

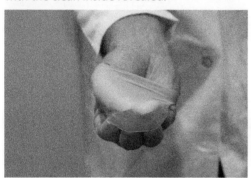 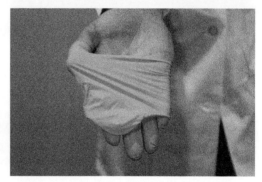

Step 10. Slide the (clean) index finger of the left hand into the gap over your right palm, then push the glove off your right hand and into the trash.

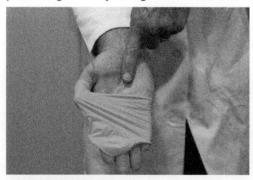 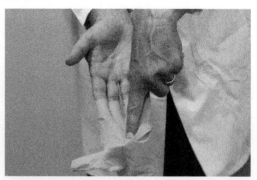

Step 11. *Wash your hands.*

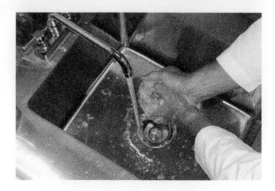

I never knew how much this procedure was loved until the nurses at Emory Healthcare told me they were "beakers" for life. This made my day and demonstrated the value in the beaking method of glove removal.

BIOSAFETY *in the First Person*

Representing the Profession of Biosafety by Ed Stygar

I don't think I can write a book about biosafety without recognizing the organization that introduced me to many who are featured in this book. Ed Stygar is the executive director of ABSA International, previously known as the American Biological Safety Association. Both he and his family have a long affiliation with ABSA International. I consider Ed a person who not only serves the profession of biosafety but advocates for the profession as well. Ed's contribution to this book tells a story of ABSA International, the members who made a difference, and how it became the organization it is today. Thank you, Ed, for your service to the profession of biosafety.

The American Biological Safety Association (ABSA) was managed by volunteers until 1991, when they decided to hire my father's company, Stygar Associates, to manage them. Stygar Associates was a company that managed nonprofit associations that were too small to staff their own office but needed professional management. I began working with ABSA in January 1993. I had just graduated from college and was thinking about going to graduate school. My father, Edward J. Stygar, Jr., asked me to work for him temporarily because he had recently experienced a sudden loss in eyesight.

In 1993, ABSA membership was about a quarter of what it is today. Some of my first tasks for ABSA were marketing and exhibit sales. At my first ABSA conference in Albuquerque, New Mexico, in 1993, we had 17 exhibitors and approximately 250 participants; since then, we have peaked at 90 exhibitors and 750 participants. Being a business major straight out of college, I had no clue what biosafety was, but I was immediately impressed by the passion and knowledge of ABSA members.

After I had worked for my father for a year, his eyesight returned, but I decided to work for him on a permanent basis. Looking back, it may not have been my ideal choice, but it is a choice that has benefited me tremendously, both professionally and personally. My father became a great mentor to me and taught me about business, doing the right thing, and helping organizations grow. I was always impressed by his knowledge, accomplishments, and ability to handle difficult situations and how he genuinely wanted to help organizations. As he was the "business guy," his advice and decisions were not always popular, but he was always thinking about the long-term health of the organizations. This is evident, as the associations that he managed, including

ABSA, had success under his supervision. We managed 10 organizations at one time, so I learned how to work with different professions, diverse needs, and unique organizational cultures. I learned lessons that could be applied to other organizations, but I also learned that all organizations are not the same. Mary Buckley managed the ABSA account at that time. She taught me to work hard and provide "service with a smile." I am very thankful for those mentorships and experiences.

The initial years of managing ABSA had their challenges. Some long-time members were not fond of having a professional management firm handling the association. We also brought a new perspective on management and growth, and it took some time for the leadership to become comfortable with this. After some success and after they got to know us, we won some of them over. As with anything, time was needed to adapt and trust new partners.

ABSA not only grew in membership in the 1990s but also added several programs for the members. There was a need for biosafety education, so the organization added preconference courses that became popular with members. There weren't any credentials for biosafety professionals, so ABSA created one, the Registered Biosafety Professional, in the early 1990s, which was mostly based on experience in biosafety. In the late 1990s, ABSA worked with the American Society for Microbiology and the National Registry of Certified Microbiologists to establish a test to credential the Certified Biological Safety Professional, which was based more on education.

Another very important decision for ABSA in the 1990s was to publish *Applied Biosafety*, the first journal focusing on biosafety. Publishing the journal was a big jump for ABSA, with the challenges of added expense and getting the right team to produce it. The ABSA Council worked hard to gain membership support, and the first issue was published in 1996. The first editor was Melvin First, followed by Richard Knudsen and Ira Salkin. Barbara Johnson and Karen Byers, who have been the coeditors since the end of 2004, have worked with the associate editor, David Gillum, and the production editor, Karen Savage, to continuously improve the content and format of the journal. The journal has been a great benefit to biosafety and has improved every year.

There have been many ABSA members who have helped ABSA move forward through the years. All had passion about the profession and were ready to lead. These members also knew that sharing and working together were necessary for ABSA and biosafety as a whole. I remember the early conferences where members were sharing tools and information, something I hadn't seen much at other conferences or associations. Many outsiders and first timers have been impressed by ABSA members' willingness to share and network with others. Not only do they share information, but they are more than willing to share their time volunteering for the organization.

At the end of the 1990s, I was still involved with marketing and exhibits but also helped with finance and budgeting. Then the opportunity to work in more strategic areas came about, which I found most interesting. As I was getting more into strategy, so was ABSA, as they developed their first long-range plan. The association was growing, so it was appropriate to set a direction so ABSA could focus on priorities. Strategic planning is like sailing across the ocean: if you don't know where you are going and don't have a plan to get there, you will most likely get lost. Not only will you get lost, but you will lose time and resources.

Some of the valuable leaders, volunteers, and mentors in the 1990s were Jerome Schmidt, Jack Keene, Mary Ann Cipriano, Lynn Harding, Byron Tepper, Barbara Johnson, Richie Fink, Manny Barbeito, Jerry Tulis, Gerald Spahn, Melvin First, Jonathan Richmond, Joe Van Houten, Debra Hunt, and Diane Fleming. There were many more, and without them ABSA wouldn't have had the foundation for what was around the corner.

My first car out of college was an Oldsmobile—it wasn't as cool as what most of my friends were driving, but I always echoed Oldsmobile's pitch: "This is not your father's Oldsmobile." Suffice it to say, when the clock struck midnight on December 31, 1999, little did we know that it was time to buckle up and that "This is not your father's Oldsmobile" rang true about the millennium and the decade to come.

In the early 1990s, most ABSA members were microbiologists from government or military. In the late 1990s, we experienced a gradual shift to younger members coming mostly from academia but also from biotech and pharma. This shift produced the next generation of ABSA members and leaders. It was scary because it was new, but exciting because of all of the new ideas and energy.

One focus and success of the new millennium was our training and education program. As a newer and growing profession, biosafety offered few learning opportunities. The ABSA Council and committees realized this and started to expand our preconference courses. With the help of many enthusiastic volunteers, the courses grew in popularity and offerings and became a great benefit to our members. We even added spring courses and developed a week-long course for those who were newer to biosafety. The week-long course, called Principles and Practices of Biosafety, was created by a team led by LouAnn Burnett. The first course was in Richmond, California, in July 2003. The course was an immediate success and has been continually sold out to date. We originally offered the course once a year, but it is now offered twice a year. The course has received great reviews and has helped recruit many new members and leaders for ABSA.

Our October 2001 conference was in New Orleans, Louisiana, a little more than a month after 9/11. There was a flurry of activity with the media, new faces, and notable absences of those working on issues related to 9/11. I was impressed by the way ABSA's leadership handled the attention and how they adapted to the need for biosecurity information and training. Due to 9/11 and the anthrax attacks, there was more attention being paid to biosafety; this increased the demand for ABSA membership and training courses and increased oversight and regulation. In the years to follow, ABSA had an influx of presentations, papers, and courses related to biosecurity and high-containment training. Due to the building boom in BSL3 labs, ABSA saw an increase in security experts, architects, and engineers. Another effect was an increase in international interest in ABSA, biosafety, and biosecurity. It was a whirlwind, but ABSA reacted well by making sure members' changing needs were met.

With all that was happening with ABSA and its environment, there was a need to ensure that we had the right structure and focus to move forward in the changing landscape. In 2004, ABSA initiated a strategic planning effort led by Stefan Wagener and Elizabeth Gilman Duane that would help ABSA move to the next level. For the first time, ABSA used an outside consultant to do an overall assessment of ABSA and its environment and help create a strategic plan. Through this effort, we adjusted our structure, defined roles, and created long-term goals that would be beneficial to ABSA for many years to come. It was not an easy process, but it was a great step for ABSA. The plan helped us focus, work together more efficiently, and serve the needs of our members more quickly.

In early 2006, my father's company lost a couple of clients, but instead of trying to rebuild the business, he decided to retire. I had had a great experience working with him and managing different types of associations, but I did not want to continue working for an association management company because it was difficult to be a for-profit that contracted with nonprofits—cultures and priorities sometimes clashed. ABSA had nearly doubled in membership since 2001 and was in a position to open its own office with its own staff. Stygar Associates had the office space, equipment, and employees to easily transition ABSA to become a self-managing association. After some number crunching and talking to the staff, we presented an option to ABSA—the association would stay in the same office with the existing office equipment and work with the same staff, which saved having to move, train staff, or equip an office. The only change was that my father was retiring, and I was the proposed executive director.

The ABSA Council deliberated at a meeting in Miami and decided to open their own office. I was excited about this new opportunity to lead a stand-alone organization, to be integrated into the organization, and to utilize my leadership skills. Not only was I very fortunate that ABSA had faith in me, but I was privileged that my team of Julie Savage, Mary Buckley, Karen Savage, and Barb Minarik were joining me at ABSA. I was nervous and yet confident that I had the experience and knowledge to be ABSA's first executive director. I made sure to keep up on association management practices through books, courses, and conferences and studied to obtain the Certified Association Executive credential. Some would think that this wasn't a big change from working for an association management firm, but it was different in that I was not a contractor any more, and I was able to focus on the mission of one organization versus several. I liked the change! I eventually became an integral part of the ABSA family, building trust, and becoming more collaborative. The transition went well, the staff-volunteer relationship became stronger, and ABSA continued to grow.

ABSA had continued success in 2006 and 2007. At our 50th Annual Conference in Nashville, we had many new international conference attendees, which was exciting for everyone. Things were going well and everyone seemed to be happy with ABSA's progress. I remember being at the conference, thinking that everything was great now but wondering what would happen if ABSA had difficult times. What if we had a sudden downturn in membership and/or low attendance at the conference? At that time, the ABSA conference accounted for more than 60% of our revenue, so if the conference had a bad year, then ABSA had a bad year. In my experience, organizations tend to be a little more relaxed with using resources when the economy is good. Running a business is much easier when there is success; it is the difficult times that test everyone.

After the Nashville conference, I emphasized the need not only to be fiscally responsible but also to make sure that we were continuing to meet the needs of our members. I worked with the treasurer and finance committee to make sure that we budgeted more conservatively. It had been nearly four years since our last strategic plan (created in 2004), so I suggested that the ABSA Council revisit the plan. We had another strategic planning session in the spring of 2008. Some of the elements of that plan were hiring a director of education and starting a task force to explore the possibility of a lab accreditation program. We now have KariAnn Deservi as our director of education and continue to drastically expand our catalogue of distance learning opportunities. We also launched the High-Containment Laboratory Accreditation program and accredited our first entity in July 2017.

ABSA had another record year in 2008, having doubled in size and resources between 2000 and 2008. Some of the volunteer leaders who were helpful during this period were Maureen Ellis, Karen Byers, Robert Hawley, Stefan Wagener, Elizabeth Gilman Duane, Richard Rebar, Joseph Kozlovac, Bill Homovec, Chris Thompson, Robert Ellis, Ben Fontes, and Glenn Funk.

Another factor that was helpful to ABSA and the biosafety community was partnerships. We partnered with many different groups for numerous different projects, both nationally and internationally. One of the longest and most beneficial partnerships has been with the Elizabeth R. Griffin Foundation. The foundation was started by Beth Griffin's parents, Caryl and Bill Griffin, after Beth passed away from the herpes B virus while working in a laboratory. Their executive director, Jim Welch, approached ABSA in 2004 about working together to promote safe science, something that is very relevant to ABSA's mission. Since the beginning, it has been a great relationship in which we have worked together on research, training, awareness, and tools for biosafety professionals. Our relationship with the Griffin Foundation also led to the partnership with Sean Kaufman at Emory University for a Leadership Institute for Biosafety Professionals, a truly unique course. The Griffin

Foundation also helped us get involved in other partnerships, such as the Global Health Security Agenda Consortium, and begin working with international organizations. Another partnership that both organizations are involved with is the USDA-ARS International Biosafety and Biocontainment Symposium, led by Joseph Kozlovac of the USDA Agricultural Research Service. The symposium filled a much-needed gap in covering biosafety and biocontainment issues in agriculture and has been well received by domestic and international participants. Other important partners for ABSA have been the Centers for Disease Control and Prevention; Association of Public Health Laboratories; American Association for Laboratory Animal Science; Campus Safety, Health, and Environmental Management Association; European Biological Safety Association; Occupational Health and Safety Administration; NIH; Sandia National Laboratories; Eagleson Institute; and the ABSA Affiliates. Partners have allowed ABSA to accomplish many goals that they couldn't reach on their own.

After 9/11 there was severe acute respiratory syndrome, the bird flu, the tsunami in the Indian Ocean, Hurricane Katrina, and then the great recession that started in late 2007. Fortunately, the ABSA Council and staff had been fiscally responsible by building financial reserves steadily for almost 20 years. With the great work of the council, finance committee, and staff, ABSA had minimal losses during the recession and came back stronger than before. Several plans were instituted to cut expenses and to add revenue streams. Some new revenues that were helpful to ABSA include the management of the USDA-ARS Symposium and the increase in distance learning courses.

From 2008 to 2010, ABSA did everything possible to keep moving forward during the recession. We had to scale down, but tried to do so without affecting members. Around 2010, I noticed that ABSA had been losing focus and trying to go in too many directions. In retrospect, our lack of focus was due to the distraction of the recession and also because the 2008 strategic plan had too many goals and objectives. I knew that with the rapidly changing environment, ABSA had to create a new strategic plan, but the real question was, how could we retain a strategic focus for a long period with rotating leadership? I did some research, talked to other association CEOs, met with consultants, read some strategy books, and benchmarked to gain more information. As a result of this research, I created and proposed a plan to the ABSA Council in which not only would we create a new strategic plan, but ABSA leadership would have annual strategy engagement to understand and efficiently carry out the plan. We would have annual Council strategy sessions and an annual strategic workshop at the ABSA Conference for all volunteers. The current ABSA president, president-elect, and executive director would also attend a CEO symposium led by the American Society of Association Executives (ASAE). The CEO symposium would introduce strategic and high-level management concepts and also help the executive committee work together more efficiently. Not only would this overall plan help the ABSA Council and staff keep a strategic focus, but it would help volunteer leaders to understand the plan and make it easier for them move into higher roles.

The ABSA Council didn't immediately accept the entire plan; they decided to test it with the ASAE CEO Symposium. LouAnn Burnett and I attended the CEO Symposium in January 2011, and we were both very impressed with the symposium and the team instructing it. We learned a lot and we even found our future strategic consultant, Paul Meyer of Tecker International. We both agreed that the symposium and the ideas we gained from it would be beneficial to ABSA. After that symposium, LouAnn became the champion of our current strategic initiative. She did a lot of research and work to convince the Council to go ahead with the plan. After the Council approved the initiative, we worked together on finding the right consultant, communicating with ABSA members, preparing for the planning session, and then launching the plan. It was a two-year process from inception to launch. The initiative wasn't easy to get started and took some convincing even after the launch. The plan really excelled after the launch, and we have many new programs, initiatives, partnerships, and engaged members because of it. The ABSA Council and volunteer leaders have stayed focused and become well versed in strategic structures and management.

One item in that plan was an international focus that led to ABSA's changing its name to ABSA International. Another notable item was a focus on increased learning opportunities, which led to many more distance learning courses that have benefited ABSA members and biosafety professionals. LouAnn Burnett helped start the strategic initiative, and Karen Byers, Barbara Fox Nellis, Paul Meechan, Marian Downing, Melissa Morland, Maureen O'Leary, Patrick Condreay, and others kept ABSA moving forward. The ABSA team of Julie Savage, Karen Savage, Diane Johnson, KariAnn DeServi, Lena Razin, and Michelle Gehrke were also very helpful behind the scenes in making the plan a great success.

ABSA has evolved and grown a great deal. There were many factors involved in this transformation, but the ones that remain constant are the engaged members and their commitment to ABSA and the field of biosafety. Having worked with many different organizations, I have determined that ABSA's members are the most engaged members that I have ever seen. They have a passion for ABSA and ensuring safe science.

chapter **15**

Safety Surveillance Programs

In the middle of Liberia during the 2014 Ebola outbreak, I walked into a hospital which appeared to be abandoned. I was surprised by this, considering that we were in the middle of a serious epidemic. Medical issues don't simply stop because epidemics occur, so where was everyone?

As I explored the hospital a bit more, I heard screaming coming from the surgical room. Of course, I believed that I was hearing a sick patient, someone fighting Ebola. But to my surprise, two nurses emerged from surgery and told me that they were in the process of delivering a baby. I asked them about resources and personal protective equipment. None were available, they told me. I asked, "How are you protecting yourselves?" They shrugged and said, "There is not much we can do. Babies still come, and we do our best to be here when they do." I asked if they were aware of anyone who was being treated for Ebola. They pointed to the other side of the hospital and said there was one patient they knew about. That was when I met Moses and Maybelline for the first time.

Moses had no personal protective equipment but would not leave the side of his wife, Maybelline, who was in the last stages of an Ebola infection. She was in pain and praying. I was moved by Moses' commitment to stay with her. He told me, "She was never afraid of her patients [who were sick with Ebola]. She was always afraid of dying alone. I will not leave her to die alone here." He stayed. Maybelline did pass away, but Moses made it out of quarantine even though he really did not have much to keep him from becoming infected.

There is more to this story. Maybelline was a delivery nurse who contracted Ebola while treating an Ebola-infected mother who was suffering a miscarriage. Like many other health care providers and laboratorians, Maybelline worked on the frontlines, interacting with infectious diseases. This causes many health care workers to become sick and give up their lives for the work they do. Health care workers focus on patients. Laboratorians focus on science. While these two professions are doing their work, who is focusing on *them* to ensure that they stay safe?

Surveillance systems are a tool that safety professionals can use to ensure the safety of the workforce in both health care and laboratory settings. We will never be able to promise a zero-risk work environment when infectious diseases are involved. However, we can implement surveillance systems to catch conditions and behaviors that can increase the risk to those who are serving, others they serve with, and those (people and pets) they return home to. Surveillance is defined as making a close observation. I believe close observations are needed for all those serving in the health care and laboratory professions.

HEALTH SURVEILLANCE PROGRAMS

The physical and mental health states of those in health care and laboratory settings are not static but ever changing. Changes in health status should not be seen as barriers to serving our profession but should be recognized so our profession can better serve us. During my work at the CDC, I was asked to participate in a public health emergency response. Prior to being deployed, employees are seen by a medical doctor to ensure that they are fit for service. During this examination, my blood pressure was abnormally high. Not only did this indicate a health issue, but also it showed that until my blood pressure was controlled, I was not fit for deployment. This simple surveillance system identified a health problem and served as the reason to quickly address it.

Health surveillance programs (HSPs) can be divided into three levels: initial, reactive, and continuous. It is imperative that HSPs occur at all three levels. Simply implementing one level leaves a large safety gap. These HSPs can be implemented with minimal costs and produce a safer workforce. The costs related to anyone in the workforce dying or getting sick can justify the existence of HSPs, which are a closer observation (surveillance) of those serving on the frontlines of infectious disease.

Initial HSPs

I remain amazed at the fact that we go through extensive interviews and other human resource challenges when hiring and onboarding someone into an organization, and yet we rarely require a medical examination to determine whether someone has a pre-existing condition that may disqualify him or her from working with or around infectious diseases. I have seen those working in facilities management receive more medical screening than nurses, doctors, and laboratorians combined. This is inappropriate. Any professional who is tasked to work with infectious agents or around patients who may be sick should be screened before being hired.

Many people have pre-existing conditions that compromise immune function or place them at increased risk of loss of life if they do get sick. Even worse, many of these people don't know they have such a condition. Initial HSPs provide the opportunity to protect organizations from preventable risk that may damage reputations and lead to loss of life and increased liability. These surveillance programs also provide the opportunity for the people being screened to learn about any unknown conditions treatment of which can lead to increased quality and quantity of life.

Having an initial HSP in place is a no-brainer to me. Leaders overseeing workplaces where there is a potential of exposure to agents that can make people sick should never allow someone to work in their organization without considering that person's health status. Not having an initial HSP is both risky and irresponsible. Close observation (surveillance) of those being hired is needed.

Reactive HSPs

When dealing with infectious disease, we really don't know if an incident is an accident until we are out of the incubation period. When something unexpected occurs—an incident—it remains just an incident until it causes harm; then it turns into an accident. Reactive HSPs respond to incidents or exposures with strategic plans aimed at minimizing the likelihood of illness among those involved in an incident. Having a reactive HSP would have made a difference in the way medical professionals responded to Beth Griffin's incident. Many outstanding professionals have since implanted strategies to ensure that what happened to Beth won't occur again, including providing staff with information cards that allow medical providers to accurately assess the risk of exposure in those working around herpes B virus.

Health care and laboratory staff reporting an incident must have a protocol to follow until it is known whether the event remains an incident or becomes an accident. Reactive HSPs provide prophylactic treatments, monitoring, and incident investigations and ensure a quick medical response if people become sick as a result of the event in question. Close observation (surveillance) of those reporting an incident is needed.

Continuous HSPs

As time passes, things change, including our health and our work environment. Stresses and chronic health conditions can develop that may impact our ability to fight infection following an exposure to an infectious disease. Health care providers and laboratorians usually don't work alone; people coming and going and multiple behaviors occurring in tight spaces increase the likelihood of contamination. Although we may be brilliant at the work we do, we cannot see the difference between what is contaminated and what is not. Our immune systems and health status provide the frontline defense needed to have a chance of remaining healthy following an exposure to an infectious agent.

Both Malcom Casadaban (see page 34) and Linda Reese (see page 22) were very experienced but lost their lives because of changes in their physical conditions that placed them at increased risk when working around their biological agents. Continuous HSPs make sure that those working with and around infectious disease remain fit to do so. Close observation (surveillance) of those working with infectious substances or around infectious patients is needed.

MEDICAL SURVEILLANCE PROGRAMS

Do we always know when an exposure occurs? Nina and Amber, the two nurses working in Texas who got sick while treating an Ebola patient, never knew when or how their exposures occurred. Kent and Nancy, the two Ebola patients treated at Emory University, could only guess at how their exposures occurred while working in Africa. Some have reported that between 70% and 80% of all laboratory-associated infections come from unknown events, meaning that they cannot be traced back to a single event that could have been labeled as a known exposure. These facts call for a strategy. Close observation (surveillance) of the day-to-day health of those working with infectious substances or around infectious patients is needed.

Anyone who works around infectious disease should have specific training about the biological agent(s) involved. This training would include knowing what the agent is, how it is spread, the incubation period, the signs and symptoms of clinical presentation, treatment options, and strategies that prevent exposure. Medical surveillance programs zoom in on the clinical presentation of the illness.

Leaders should ask for and expect the immediate reporting of symptoms that match the clinical presentation of the agent their staff are working with or around. This is exactly what both Nina and Amber did while treating an Ebola-infected patient: both immediately reported fevers, minimizing the risk to others and increasing the likelihood of their surviving an Ebola infection.

In the years that I have been making this recommendation, I have received a tremendous amount of resistance. Why should I report common cold symptoms? You really want me to tell you I have diarrhea? Don't you think it is more than likely that I got this outside the laboratory? All of these questions miss the point. It is not about you. It is not about me. It is about us. Medical surveillance programs identify problems in the environment that are causing those working to get sick. When we see someone with symptoms, we report it and keep going while monitoring that person's symptoms and recovery. But if two or more people begin to show symptoms within the same environment, we may have a common-source exposure and should begin looking at the possibility of an environmental exposure.

The cost of this program is a notebook, a phone number, and a leader. Any symptom is called in, reported, and logged. When two or more from the same environment report common symptoms within the incubation period of the disease caused by the biological agent(s) being worked with, that signals to begin looking at potential safety problems within the environment.

INCIDENT SURVEILLANCE PROGRAMS

An event or occurrence that poses or has a chance to pose a risk to health and safety is an incident and should be reported. The surveillance programs mentioned above are greatly needed, but in my travels, I rarely see them in organizations. Most organizations have incident surveillance programs, which consist of making close observations (surveillance) of reported incidents within their organization. But doing only this fails to create a culture within the organization that ensures the incident surveillance programs are effective at preventing and responding to unexpected events posing a risk to health and safety. The following describes the three major barriers to effective incident surveillance programs that I find in place when visiting laboratories and clinics: the reporting process, the "blame game," and punishment of reporters.

Reporting Process

There are several reasons for reporting incidents. First, we want to be able to prevent them in future. Second, we want to be able to respond to them. Third, they create a liability. But what happens when the potential liability produces a process that hinders the reporting of incidents?

Having talked to many who serve in both hospitals and laboratories, I find that there appears to be a barrier of resistance to reporting incidents that is difficult to overcome. There are forms and interviews, plus the time these take away from regular duties, all of which serve to dampen the likelihood of an incident being reported, especially when it only seemed like "a simple incident."

During the time I served in the Emory Healthcare Isolation Unit, we had an incident reporting system that I grew to love. We developed a culture of family and caring, so no one was embarrassed to report an incident and, further, it was embarrassing to fail to report one. Date, time, incident, person, and response were the items that were logged and then discussed during a family-style meeting, which occurred on a daily basis. It was simple, it was nonjudgmental, and it worked. No doubt, long reports may have to be written, but please ensure that your incident reporting system isn't a barrier to incident reporting. That turns it into a reverse safety issue.

The Blame Game

Many incidents aren't reported because the focus is not on *what* went wrong but on *who* did wrong. When incident reporting becomes a blame game, people simply stop reporting. Remember, incident reporting asks people to report a potential mistake. Why would an employee report something that could get them or others in trouble and place them in jeopardy of losing their job, facing professional scrutiny, and tarnishing reputations that have taken years to establish?

Organizations can improve incident reporting and transparency if the *who* is replaced by the *what*. People are fallible, and if one person makes a mistake, is it not likely that another may come along and make the same mistake? If organizations focus on what caused the mistake to happen rather than who made the mistake, they can develop a solution that protects all those who do the *what*.

Punishment of Reporters

Sometimes, we punish behavior intentionally. If organizations intentionally punish those who report incidents, those incidents will not pose the greatest risk; the culture of the organization will. However, there is also the unintended punishment, or in other words, having processes that stop the reporting of incidents unintentionally. These could be a reporting system that takes too long to complete, or even offering a reward when there are zero incidents. Can you imagine being the one to report an incident and thus being the reason that nobody gets a reward?

In order for incidents to be reported, staff must be encouraged to do so and be reminded of why it's important to report. Reporting processes must be quick and simple and focus on what went wrong. If people feel as though they will be judged, scrutinized, or labeled as a result of the incident, there is a greater likelihood that an incident will go unreported.

VIDEO SURVEILLANCE PROGRAMS

It was my first football game, and I was kicking off. I positioned the football on the tee, ran back behind the kickoff line, lifted my arm, dropped it, ran toward the ball, and kicked it down the field. As I ran through the kickoff, a defensive lineman hit me unexpectedly, slamming me to the ground. When Monday came around, in front of the entire football team, our coach played video that showed that hit over and over again. Even though this elicited jeers, smiles, and laughs from my teammates, the coach was making a point. He used the video not to punish or criticize me but to teach me. He showed me that my feet were pointed in the wrong direction and that I was looking down at the ground rather than up at the field. These observations improved me and improved my safety on the field.

I have never completely understood the apprehension staff express when video cameras are placed in the laboratory or hospital setting. Standard operating procedures are written and expectations are established, and yet the programs developed to ensure compliance with these expectations are weak, if they even exist at all. The concept that "big brother is watching" is not a bad one unless "big brother" punishes you when you make a mistake. Video surveillance programs are needed because close observation (surveillance) of the day-to-day behavior of those working with infectious substances or around infectious patients is needed.

Coming from a background of playing sports, I can tell you that there were times when I would argue that I was not doing something until the tape revealed that I was doing it. Video does not lie. It tells the truth about behavior and bridges the gaps between what people think and say they do and what they really do.

Video cameras should not be used to punish people, unless they are engaged in criminal or unethical behavior. Criminal and unethical behaviors place the organization and staff at risk and these behaviors must be punished (stopped) for the safety of the organization and those working in the organization. Beyond this, video surveillance should be used to teach, mentor, coach, and correct behavior deficiencies. Video surveillance programs provide opportunities to improve safer behaviors, compliance, and workforce accountability. I have witnessed video making vast differences in how people behave, especially when it is used to support them and reinforce proper behaviors versus punish them. If you implement a video surveillance system, be sure to use it to improve rather than destroy those on camera.

INVENTORY SURVEILLANCE PROGRAMS

When I was little and bound to a "two cookies for dessert" rule, I would take any opportunity I had to sneak an extra cookie. My little brother would do the same, especially when our parents left us alone at home. My parents were quick to catch onto this, so before they would leave, my mom would check the cookie jar to see what was in there. She would not count the cookies but just make a quick visual inspection. After she would leave, I would look in the jar, and if there were lots of cookies, both my brother and I would take one. Mom was always fair; she would not accuse us unless she knew for sure a cookie was taken. So, if there were only two or three cookies left, I would not dare touch them because I knew that she knew exactly how many cookies were in the jar. My point is, that if you don't know what you have, how can you tell what you are missing? Close observation (surveillance) of the biological inventory is greatly needed.

Inventory surveillance matters because it is not only a safety issue but a security issue as well. Inventory records are different from inventory surveillance programs. If a bank statement is an inventory, an audit is an inventory surveillance program. Take a close look at your inventory—who has access, who is accessing, deposits, withdrawals, activity locations, transfers, inactivation of agents, and justifications are all critical components of an inventory. Ensuring that these are being logged accurately is a component of an audit.

If you want to test your inventory program, take something from where it should be and hide it somewhere else. Ask the staff to facilitate an inventory check and see if you are notified of the loss. If you are, it is a successful audit. If you are not notified of the loss or the loss is reported as an inactivation, then there is a failure in your surveillance program. Such a failure should be immediately addressed to ensure honestly, integrity, and transparency at the highest levels.

CONCLUSION

Safety surveillance programs are essential. Making close observations of who is doing the work, how they are doing it, what is being worked with, and the incidents during work is a common-sense approach to safety. Some see surveillance programs as intrusive, but I see them as imperative. The belief that we know more than what we really know can lead to our hurting ourselves and others. Surveillance systems provide the observations needed to ensure safety within organizations.

 BIOSAFETY *in the First Person*

Experience in Years by Mike Pentella

Mike Pentella first introduced me to the challenges that separate research and public health laboratories. Extremely kind, very wise, and an exceptional teacher, Mike goes out of his way to share his experiences with others. Mike advocates strongly for public health laboratories, and we should listen to him. We are potentially one outbreak away from destroying an under-resourced, -valued, and -staffed necessity! Public health laboratories need a great deal, including safety training. Mike, along with the Association of Public Health Laboratories (APHL), continues to serve and advocate for more resources. He is an extraordinary person and professional.

EARLY YEARS

In January 1973, as a recent graduate with a Bachelor of Science degree in microbiology from The Ohio State University, I got a job in a hospital clinical microbiology laboratory. I soon learned that working with human specimens was nothing like my lab experience in school. My first day started with specimen setup the first hour that I was there. There was no orientation or safety training. What surprised me was that the person training me was drinking coffee and smoking a cigarette in the lab. Fortunately, the lab had recently hired John Thomas, a Ph.D. clinical microbiologist who was doing his best to change things. The lab got much better in the months ahead. We wore lab coats that we purchased ourselves and took home to be washed at our own discretion. I used the same lab coat for a whole week unless I spilled something on it. The infection control nurse had her office down the hall from the lab. She mentioned handwashing to us a few times, but nothing else was ever said.

The lab wanted to bring tuberculosis (TB) testing in-house and sent me to a local TB sanatorium to be trained by their laboratorians. The TB sanatorium had an excellent lab and followed safety precautions very well. They taught me how to use a biosafety cabinet. They must have done a pretty good job because I never seroconverted on my purified protein derivative after thousands of specimens processed over the years.

After a couple of years, I left the clinical lab for graduate school. During my graduate school career, no one talked to me about biosafety. I was working with *Chlamydia trachomatis*. I used a biosafety cabinet for all my work. Biosafety was not covered in any of my graduate coursework. I supported myself by working part-time in a clinical microbiology lab. There, too, biosafety was never mentioned.

CLINICAL LAB

After graduate school, I returned to the clinical lab because I found the work more satisfying in terms of helping patients. It was around this time that our hospital first began to see a few patients with AIDS. Through attending the American Society for Microbiology annual meetings, I started to learn more about the disease and testing for it. I came back and shared the information with the medical technologists in the lab. There started to be a real concern for the risk the staff was taking by handling these specimens. Our hospital, through the infection control committee, took every step possible to minimize the risk and followed the CDC recommendations. Even so, staff felt a great concern about working with specimens.

We had a practice on the blood culture bench of cutting used needles after performing subcultures. There was always splash and splatter of blood on and around the cutting device. I recognized that this was creating a risk for us and stopped the practice. We stopped mouth pipetting and recapping needles and switched from glass to plastic whenever we could. Still, staff feared that they would have an exposure, and some spoke of wanting to leave the field because of it.

At that time, the specimens were labeled with a sticker as "HBV" or "HAV" for hepatitis B virus or hepatitis A virus. It seems ridiculous, but some people believed back then that you could tell by looking at the patient whether he or she was infected with HIV. After a few years, the hospital adopted universal precautions, in which every specimen is considered potentially infected with a blood-borne pathogen (HIV, HAV, HBV, and HCV). This was a very important first step toward preventing infections. Because I was managing infection control as well as the lab, I worked on developing education and training for nursing and other allied health professionals about blood-borne pathogens. It was the start of my interest in biosafety. Since then, offering training in biosafety has been one of my areas of interest. At the hospital, we started with in-person lectures but then developed a series of three movies so that everyone received the same information. These projects interested me because I learned more about the principles of adult learning. We tried to make the movies entertaining. We followed the same characters through the series of videos, from drawing blood and getting stuck with a dirty needle in the first video, through postexposure testing and acquiring the infection in the second video, to the death due to infection of the exposed health care worker in the third. The soap opera-like approach engaged the staff and facilitated learning.

Vaccines for Lab Staff

My interest in infection control and prevention strategies led me to realize that biosafety in the lab is important. Eventually, I went on to manage both the infection control department and the clinical microbiology lab along with other lab sections.

In the mid-1980s, the HBV vaccine became available. I had read that laboratorians were at high risk for acquiring HBV. I was excited when the infection control committee included the laboratory staff in the recommendation to receive the vaccine at no cost to the employee, because at that time it was not an Occupational Safety and Health Administration (OSHA) requirement. Some of the staff feared the vaccine because it was prepared from sera from men who have sex with men and could contain HIV that was not killed in the vaccine preparation process. There was no evidence for this potential risk, but there was a great deal of discussion about it in the lab. The hospital saw a high number of patients with HBV. I weighed the risk of a possibly defective vaccine against the risk of HBV and decided that the risk of getting HBV was much higher than the risk of getting HIV from the vaccine. I shared my reasoning with the staff, and about half of the staff decided to get the vaccine. There was never a documented case of HIV from the vaccine. Later, getting the HBV vaccine would become the norm for lab staff.

Lab-Acquired Infection

No one working with me had ever acquired an infection in the lab until one of the microbiology medical technologists became infected with *Shigella*. While concerned for her well-being, I was also very surprised because this med tech was very good about handwashing. She would never go to break or lunch without handwashing and frequently washed her hands during the workday. The sink was right next to the stool bench to make it convenient. Working with staff, we hypothesized that the infection might have been acquired from contaminated surfaces, and we decided that, in general, we were not good about disinfecting environmental surfaces. We postulated that *Shigella* might have contaminated the bench top or the phone or something else that was commonly touched. After that discussion, we changed our protocols so that surfaces were routinely disinfected each day. However, I often observed that the procedure was not being adequately performed, because the practice was to spray down the surface with a quaternary ammonium compound and let it air dry. Just spraying the surface did not cover the entire work area evenly; it needed to be wiped down. Through the years, I tried to get us to improve the housekeeping practices, but it was difficult because the workload was heavy and everyone was in a rush to leave at the end of the shift.

Tuberculosis Exposures

Our biggest problem in that lab was exposure to TB. We had several instances where there were skin test conversions. The first one was in a morgue attendant, or diener, in the morgue. The diener was never required to wear respiratory protection when the pathologist sawed into the sternum. The pathologist did advise the diener to turn his head during the sawing process. The pathologist was adamant that the diener's TB exposure was from the community and not associated with working in the morgue. When the worker's compensation investigator came, he first questioned the pathologist and then questioned me as the manager. I reviewed the data of positive cases of TB from the morgue with the investigator and explained that the hospital did not require the diener to wear a mask. The investigator determined this to be the most likely cause for the infection and made the case a worker's compensation claim. At that time, the source of the exposure could not be proven, but the incident did serve to change requirements so that the diener had to wear a mask during that procedure. That was before the realization that we should be wearing N95 respirators when TB was a risk.

Biosafety Cabinets

We had several microbiology medical technologists with skin test conversions as well. The biosafety cabinet was certified every six months, but we never trained anyone on the proper use of a biosafety cabinet. In retrospect, improper use of the cabinet probably led to some of the exposures. Because of the exposures, I decided not to permit anyone who was pregnant or undergoing chemotherapy to work in the mycobacterium laboratory. I could not do that today because of human resource rules, but in the 1990s, it was an acceptable practice.

I wish I had known more about the risk of lab-acquired infections while managing the clinical lab, because some of our practices placed staff at more risk of exposure than was necessary. During my time there, it was not unusual for us to work with *Brucella* on the open bench. I recall several cases of *Brucella* isolation, and all were handled on the open bench. We were very fortunate that no lab-acquired infections resulted from those experiences. The same can be said for working with *Neisseria meningitidis* on the open bench. In that lab, we isolated *N. meningitidis* from cerebrospinal fluid or blood cultures five or six times a year. In the late 1990s, I became aware of the recommendation that microbiology technologists be vaccinated for *N. meningitidis*. I took the recommendation to the infection control committee, and they agreed to offer the vaccine to the microbiology staff; however, very few of the staff took advantage of the offer. It was not mandated. We did not move to working with the organism in the biosafety cabinet, which we should have. At that time, because of my lack of knowledge, I did not perform a risk assessment that would have led to a better understanding of the risk of exposure.

Staff Training in Biosafety

In the clinical lab, we did not provide much training regarding biosafety. Before new staff members started in the lab, they would hear about vaccines and receive blood-borne pathogen training during their hospital orientation from employee health staff. On a new employee's first day, I would point out the location of the fire extinguisher, what to do if there was a fire drill, the location of the eye washing stations, and the sinks for handwashing. That was all the information that I passed on to them. Looking back on it, I recognize how inadequate this training was. It was more than I received but totally insufficient by today's standards.

In general, the requirement of our accrediting bodies for biosafety standards is very. In the clinical lab, we were College of American Pathologists accredited. The inspectors never asked us about biosafety practices in the lab. This was unfortunate, because it was a lost opportunity to make improvements in this area.

PUBLIC HEALTH LABORATORY

After the 2001 anthrax attacks, I decided that I really wanted to work in the public health lab so that I could be on the frontline of the battle. I found a new position at the Iowa State Hygienic Laboratory. Biosafety in the state public health laboratory was a higher priority than in the clinical lab because of the great risk of exposure to emerging pathogens and established pathogens. However, if you look at the lab-acquired infection data, it is obvious that most exposures occur in the biosafety level 2 (BSL2) environment. From that perspective, a clinical lab is just as dangerous.

The first lab-acquired infection occurred because of a laceration with a scalpel blade while someone was working with dead crows being tested for West Nile virus. That incident was reported in CDC's *Morbidity and Mortality Weekly Report* (*MMWR*). Another lab area that had a very high risk was the rabies lab. You could enter that lab only if you had been vaccinated for rabies, and even though I was not performing testing in that lab, I got the vaccine so that I could enter the lab safely and consult on testing when needed.

Around 2004, I started having discussions about biosafety with our laboratory director, Mary Gilchrist. The public health lab was designing a new lab, and biosafety practices were an important consideration of the design process. One of my responsibilities was the select agent program for the lab, and consequently, I spent a great deal of time establishing our biosafety program for working in the BSL3 laboratory. Dr. Gilchrist had authored a chapter in ASM's reference *Biological Safety: Principles and Practices*, and when the next edition of the book was proposed, she asked if I wanted to coauthor the chapter with her. That was a great opportunity to study the topic of biosafety in more depth. I also thought that biosafety would be a good topic for continuing education. I was giving talks throughout Iowa to clinical laboratorians and included biosafety in those sessions. The talks were well received, and I decided that I would propose a session on biosafety for the ASM annual meeting. Because many laboratories were then dealing with select agents, this was an important topic to address at the meeting. I also attended several meetings that explored biosafety in more depth. One in particular, sponsored by Sandia National Laboratories in New Mexico, went into great depth regarding biosafety risk assessments. This was very enlightening for me because it created a framework to examine biosafety practices. I became a big fan of biosafety risk assessments. I started incorporating risk assessments into my work and biosafety presentations.

Biosafety is a difficult area to study because there is no requirement to report exposures or infections. OSHA requires employers to report fatalities, but because most of the exposures are to pathogens for which there is treatment and from which the exposed and infected individual will recover, most exposures do not end in fatalities. Also important is that the exposed individual frequently experiences stress from that exposure and the concern for serious complications or sequelae if an infection occurs. One person I worked with in the public health lab was a prime example of this. He experienced three exposures to a select agent. Each time, the exposure was recognized, and he was placed on antibiotics prophylactically. He said that each time he was concerned about what would happen if he did become ill and how his family would deal with it. He worried about exposing his children to the pathogen. These were serious concerns for him, and eventually, he left the lab to work in a different field. He was a knowledgeable and talented public health microbiologist, and his departure was a real loss for the lab. Some individuals have published accounts of exposures that occurred, and those are helpful in preventing exposures in the same circumstances. If someone is exposed to a select agent, then the CDC has that information. But the number of exposures related to select agents is very low. Naturally, laboratories whose staff experience exposures from non-select agent pathogens do not want these events to become common knowledge. It would make recruitment of new staff difficult and blemish the facility's reputation. However, the lack of data hurts our study of prevention and mitigation. It is well recognized that lab-acquired infections are underreported. The data available from published studies are old, and facilities are reluctant to report the accidents that occur.

Brucella exposure continues to be a major risk in the clinical microbiology lab. Not only the person working on the culture but usually everyone in the room is considered to be exposed. If an exposure occurs, check the CDC web page for the latest recommendations on prophylaxis. What makes *Brucella* so difficult is that it sometimes looks like a Gram-positive coccus on Gram stains of broth cultures. This is sometimes the case with *Brucella* in a blood culture. The microbiologist is then confused by the culture results and can identify the organism as a Gram-positive coccus when he or she is really dealing with a slow-growing Gram-negative coccobacillus. This happened a few years ago at one of the clinical labs in Iowa. I asked he lab to send me the original blood culture Gram stain so that I could review it. I looked it over and could definitely see why they called it a Gram-positive coccus. I always encourage clinical laboratories to not hesitate to use their biosafety cabinets when they have any question that the sample they are working with may be *Brucella*.

COMPETENCIES BUILD STAFF CAPABILITIES

Around 2009, APHL and CDC convened a group to work on biosafety competencies. It was an honor to be invited to participate in this effort. I was very excited about this because of my experience in working on competencies in the hospital environment. I had seen how, in the hospital environment, competencies had improved practices and focused the attention of the staff on best practices. After about two years of work, the document was published as an *MMWR* supplement. Unfortunately, biosafety competencies have not been incorporated by many labs. If we regularly monitor competency for the performance of a test, why are we not doing so for biosafety? The most common answer I get from the clinical laboratorians is that there is not sufficient time for monitoring and there is no incentive to get it done.

EBOLA IN THE UNITED STATES

We had the same response from laboratorians when they were faced with working with specimens from patients suspected of having Ebola that we had seen in the 1980s with HIV. There was fear, and patient testing sometimes suffered. This is not the way we as professional laboratorians should respond to an emerging infectious disease. We know there is a risk. We do a risk assessment and then we mitigate the risk.

In Massachusetts, we very early on guided our clinical labs to do a risk assessment. We sent out a risk assessment template with examples. There was still fear, but I believe that if risk assessments were common practice for clinical laboratories the fear would diminish. An important outcome from the Ebola experience in 2014 for clinical laboratorians should be to know how to face the risk of an emerging infectious agent—that is, know how to do a risk assessment and take mitigation measures.

CDC and APHL recognized the problems faced by public health and clinical labs in 2014 during the Ebola epidemic. As part of the response to the problem, APHL formed the Biosafety and Biosecurity Committee. I was asked to chair the committee and agreed to do so in the hope that the work of the committee could make a big difference in how we face an emerging pathogen in the future. This APHL committee is composed of members from public health, clinical labs, and CDC. Its purpose is to support biosafety and biosecurity efforts on a national level. In the more than four years since it was formed, this committee has made many contributions to biosafety practices. Many of the products of this committee are available at no cost and can be found at https://www.aphl.org/programs/preparedness/Biosafety-and-Biosecurity/Pages/default.aspx. These products include a biosafety checklist, an Ebola risk assessment template, and other resources. It has been a pleasure for me to work with my colleagues on this committee, and I have learned a great deal from them.

One of the products of APHL's Biosafety and Biosecurity Committee is the position statement that urges all labs to improve biosafety practices. It calls for all labs to perform a risk assessment. An effective biosafety program has risk assessment at its core. Then mitigation steps, competencies, and training follow until a culture of safety is established in the facility.

ESTABLISHING A BIOSAFETY OFFICER POSITION IN PUBLIC HEALTH LABS

Funding to support a biosafety officer in every public health laboratory came through the CDC Epidemiology and Laboratory Capacity grant (2014). This grant is available to all state and some city public health laboratories. This was especially helpful because many public health laboratories did not have a biosafety officer at that time. The expectation was that biosafety officers would first resolve biosafety issues in their public health laboratories and then reach out to clinical laboratories in their jurisdictions and help with their biosafety problems.

The biosafety officer needs to be an expert in microbiology, safety, leadership, and training. There are not many people who have all of these skills. Therefore, recruitment became a problem because there were not sufficient numbers of people with the experience and background needed to fill these positions. Eventually, all the positions were filled, but the new hires needed more training. Consequently, APHL established a training program to develop skills in the biosafety officers. It was my pleasure to serve as one of the trainers for these sessions. The training sessions for the biosafety officers helped them build the competencies that they needed to fill their role and responsibilities. An outcome of biosafety officers coming together is that they formed a community of practice. I have been very impressed with the biosafety officers and enjoy working with them. They have become an excellent resource for clinical labs and have accomplished a great deal to improve biosafety practices in the United States.

THE PRESENT

My work and interest in biosafety continue. I write and provide talks when asked. I have a great passion to see clinical laboratories put strong biosafety practices in place to protect their staff and for the next time we face an emerging pathogen. I hope that the response will be an educated approach starting with the risk assessment. I hope to see funding made available to perform studies that will allow us to offer the most effective mitigation of risk. I am concerned that, because we want to avoid an exposure at all costs, we sometimes overreact and use too much mitigation. By designing good studies, we can better determine actual risk and avoid the tendency to overprotect. Incurring some risk of occupationally acquired infection is necessary for working in a clinical laboratory. How much risk is reasonable to accept is a question that we must work together to answer.

chapter 16

Responsible Leadership

One challenge of being a safety consultant is that, as I am asked to visit and serve the needs of an organization, I am limited to what is shown to me. In my experience, what an organization says it does and what an organization actually does can be two very different things. The gap between what is said and what is done may be directly related to overall level of risk specific to safety within an organization.

Let me draw an analogy using family rules. Occasionally, my wife and I will leave our children (16, 15, and 12) at home and enjoy an evening out for just the two of us. But we were suspicious that things weren't going as well as desired when the children were unsupervised. My wife and I decided to see what was actually happening on our date nights, so we installed a Nest system in the kitchen and living room. The Nest system includes a camera that you can put almost anywhere to allow you to watch what is happening from your cell phone or any other device connected to the Internet. Of course, we did not tell our children we put the camera up.

In preparation for date night, I asked each of our three children—separately—how they behave when we were gone, and I was not surprised to hear that all of the family rules were being followed. Later in the day, my wife and I got ready, left for our date, and drove out of sight.

Within 15 minutes of our departure, behavior broke down. It started with name calling, went to some roughhousing, and escalated to indoor baseball mixed with keep-away, followed by a serving of soda and cookies. At that point, I turned the car around, parked in the driveway, and opened the garage door (which is right by the kitchen). We watched as the kids sprang into action, throwing things away and cleaning up the house faster than I had ever witnessed. We delayed our entrance to allow them to get things in order…does any of this sound familiar?

This doesn't just happen to parents. It also happens when safety professionals come in and, when seen, produce a response among those in the organization to "act safe" rather than behave normally. However, acting safe does not make you safe; safety is something that you are, not something that you temporarily do.

This illustrates the difference between a safety culture and a safety climate. Using the iceberg analogy, what you see above the waterline is the safety climate. Safety climate includes standard operating procedures (SOPs), what laboratory staff say they do, and what leaders, regulatory officials, and safety officers believe is being done within laboratory environments. What you cannot see below the waterline is the safety culture. Safety culture is what really happens within all aspects of the laboratory when nobody is looking.

This is the last chapter of this book for a very important reason. In fact, if you want leadership to engage, please consider having them read this chapter, because it is a call to all leaders of all laboratories not only to engage but to become a real member of the safety team. To do this quickly and efficiently, I will present my top 10 leadership recommendations, which when applied together will produce a theme of responsible leadership.

1. LEADERS PREPARE STAFF APPROPRIATELY

"Who wants to go to safety training?" If asked that question, I doubt many laboratory staff would raise their hands and jump at the opportunity! To be honest, they have a good reason. Many laboratory staff have been forced to attend training programs that have the sole focus of checking a regulatory box rather than developing them as safety professionals. Yes, leadership should be preparing laboratory staff by training them to be safety professionals. We rarely have the resources to always position a safety official over the shoulder of laboratory staff to watch over them. So, safety must not be done *to* scientists, it must be done *with* them and *by* them.

If leaders do not encourage laboratory staff to train, who will? Even if leaders support training, how much training should be supported? Believe it or not, it depends. In doing a risk assessment for training purposes, I believe leaders should look at the hazards that each staff member encounters in the laboratory and then ensure adequate preparedness among all laboratory staff, fairly. I will be discussing the concepts of "fair" and "equal" later in this chapter, but the fact is that we spend more time and resources training laboratory staff who sit in a controlled environment at a biosafety cabinet than we do with those who are working with animals that may be sick or not feeling well and may unpredictably act aggressively toward animal care staff.

Leaders must understand the hazards that their laboratory staff encounter and assign preparedness strategies accordingly. I believe that, depending on overall risk, laboratory staff training should account for between 3% and 5% of on-the-job time. If someone works 40 hours for 50 weeks of the year, that is 2,000 hours of on-the-job time. Depending on what they are doing, laboratory staff should be spending between 60 and 100 hours in training to be similar to other industries sharing similar risks, which include the airline, police, fire, rescue, and military professions.

Laboratory staff who work in some of the most dangerous environments spend less than 1% of their work time in safety training, and that is simply not enough to prepare or control human risk factors that laboratory staff encounter. In short, leaders should advocate for good training and lots of it.

One of the surest signs of a leader is the acceptance of responsibility when something goes wrong. Leaders own their own failure and credit others for their success. Because leaders usually write the paychecks for safety, they should also ensure that the four primary controls of biosafety are implemented responsibly. Useless engineering controls can cost millions of dollars to purchase and thousands of dollars to maintain, with little to no justification or evidence that safety is increased as a result of such a purchase. This does not only happen in engineering, it happens with personal protective equipment (PPE). Laboratory staff are sometimes asked to wear too little PPE, but most of the time they are given PPE that is uncomfortable and impractical. Leaders should care about this issue, because if something is uncomfortable, chances are it is not going to be used, or if it is used, it will not be used properly. Even if we have good engineering and PPE, if we don't ensure that good SOPs are written, failures in safety will occur regardless of the amount of money we spend on engineering and PPE controls.

Leaders must prepare laboratory staff by providing a well-designed laboratory, functioning laboratory equipment, comfortable PPE, and SOPs that have been evaluated, validated, and verified. Leaders must work with laboratory staff and safety officials to provide adequate protection and ensure that a risk assessment for protection of laboratory staff is done.

2. LEADERS ARE FAIR BUT NOT EQUAL

I realize that the words "fair" and "equal" are loaded in our society. However, leaders must understand for the sake of leading that they must treat all staff fairly but cannot treat them equally.

During a past work experience, I witnessed something that still astonishes me. Within my division, there were individuals who showed up to work and worked very hard and long hours. However, there was one employee who did not. In fact, any time someone asked this person to do something, he would complain, argue, delay, and, if pressured, file a complaint. It got to the point where this person would go into his office at the start of the day, close the door, and leave at the end of the day. Nobody knew what he did, and to be honest, nobody really cared. When those who worked hard requested a new detail (special assignment), they were turned down. However, when this person, who never worked and avoided involvement at all costs, made the same request, it was granted. When we asked the leader why this was the case, we were told that the organization could not afford to lose us (the hard workers) but that the loss of this person during an emergency really wouldn't matter.

One day I noticed this person leaving the leader's office. I asked the leader if everything was okay. The leader was excited and told me that the person who had done nothing for over two years had just asked for a recommendation to be promoted to another center. Confused, I asked the leader, "Why are you excited?" Keep in mind, I was under the impression there was no way this leader would recommend this individual for a promotion. Instead, the leader said that he would write the recommendation letter today with the hope that the person would be gone by the end of the month. Shocked, I said, "You are going to recommend this person?" The leader explained that by removing this person from our team, it would open a position and allow our team to select someone who would work and ease our collective burden. I was astounded. He was promoting someone who had done nothing! What a dysfunctional process! When I tell this story to many governmental organizations and people laugh and relate to it, this means it is still happening today, which is even sadder.

All staff should be treated fairly, which is in accordance with the rules and standards of the organization. Leaders must make sure that all staff have access to resources needed to succeed. Leaders must be fair in all ways, making sure that no single staff member is deprived of what an organization promises to provide for the success of both the employee and organization. However, we are not dealing with a computer or machine, we are dealing with humans.

When humans are given resources, some choose to work hard and some don't. If leaders decide to treat everyone equally, regardless of how staff choose to work, perform, and apply their skills, it will certainly demotivate critical staff, kill morale, and fail to retain the best workers in the industry. Leaders must treat all staff fairly but cannot afford to treat everyone equally. Blending these two and saying they are the same is a huge leadership blunder.

3. LEADERS PROTECT ALL STAFF

When an individual is hired, leaders must accept him or her as part of the family. Mistakes will occur in the laboratory. When they do, leaders have to decide whether to support the staff or turn their back on them. There is no gray area here. Many leaders will play the blame game, focusing on *who* rather than *what* went wrong.

However, I propose that leaders consider a different strategy. When an organization hires, it is hiring a human, and humans, no matter who they are or how much experience they have, will in time make a mistake or two, or three, or even more. A person who makes a mistake is not incompetent; that is different. If the leader hires someone who is incompetent, then the incompetence is the burden of the leader, and there probably needs to be a different screening and hiring process. Is it fair to ask a dolphin to climb a tree or a monkey to swim like a dolphin?

Leaders make sure to hire people with the right skills so people aren't set up to fail.

When staff make a mistake while they are doing their best to follow the expectations put in place by the organization, it is imperative that leaders protect the staff from negative consequence. Staff should not be scrutized for being human. Mistakes happen; this is part of being human and a risk we must accept. However, if the laboratory staff know what is expected of them, are adequately prepared, and intentionally choose not to comply with organizational expectations, then for the safety of themselves and those around them and the reputation of the organization, the leader must immediately relieve them of their laboratory access and duties.

I do not believe in removing laboratory staff from the work environment when mistakes happen. In fact, I feel quite the contrary. When someone makes a mistake or presents an opportunity to identify the need for any kind of improvement, I believe this person should do the action over and repeatedly until mastery is demonstrated. I learned this at the age of 10. For some reason I had this wonderful idea that if I did not do the dishes correctly, although I might get in trouble the first time, my parents would never ask me to do dishes again. When it was my turn to do the dishes, I decided I would implement my plan. I didn't scrub and didn't wash very well. When I left the kitchen, my mom noticed this and asked me if I had finished the dishes. I told her yes and watched her walk into the kitchen. I remember vividly hearing her sigh and calling my dad into the kitchen. I watched my dad look at the dishes and expected him to begin punishing me. To my surprise, he simply left the house.

About 30 minutes later my dad returned and called me into the kitchen. According to my memory, here is what was said.

"Sean, can you please come in here?"

"Yes, Dad," pretending to not even know what was going on.

"Son, it looks like we had a problem with dishes today."

"Yes, Dad. I tried really hard."

"Well, son, I am not going to debate whether you tried or not. Fact is they did not get done."

"I guess I am not very good at dishes, Dad."

"I figured you would say that. So, I just visited our neighbors and asked them if they would be willing to let you come and do their dishes this evening. All of them agreed. You will start at their homes and when done return to our home and demonstrate that you can do the dishes. Practice makes perfect, son."

When we hire human beings, they have baggage and habits that they have carried into the workplace. Leaders address these pre-existing conditions with care and trust. We don't quit on laboratory staff, certainly not staff who are trying. We commit ourselves to their development and teach them that they can do things they themselves do not believe they can do. If, however, someone thinks that he or she is more important than the safety of others or the reputation of the organization, leaders have no other solution than to let that person go.

As a volunteer, I have facilitated counseling sessions for parents who are losing a child to drugs. It is very hard. The child has become violent, steals from the house, and is hurting the family. The parents see what looks like their child, and the child acts in certain ways to prey on the emotions of the parents. The parents never want to give up on the child, but in order to save themselves, their family, and their home, they must. The decision of the parent to stop enabling such destructive behavior is a difficult one to make. I have seen top organizational leaders struggle with this same issue. They have a scientist who brings millions of dollars into their institution yet is choosing to take unnecessary risks and thereby placing their staff and reputation at risk. Rather than addressing the misbehavior, they tell the safety officer to ease off, look the other way, and allow the person to get away with behavior that places science at risk. This choice is a mistake. Leaders protect staff who are trying but they get rid of staff who believe they don't need to even try to follow the expectations.

4. LEADERS PROMOTE SCIENCE AND SAFETY

Just for a moment, imagine that organizations that house laboratories have a family structure. Both leadership and safety represent the parents. Staff scientists in this analogy represent the children, not because they act like children but because children are the legacy of the parents and the parents are doing everything they can for the children. Without the children, there are no parents. Both leadership and safety are there to serve the scientists and their science.

I use this analogy because I want you to consider what happens when one parent belittles another parent in front of the children. For example, one parent says to the children, "You don't have to listen to the other parent. Just do what you want to do. I have your back." If this happens, it kills the structure of the family. When leaders and safety officials do not promote each other for the sake of science, it begins to erode the culture of safety within an organization by splintering leadership and safety efforts.

I like to think of safety as the family dining table and safety training as the food that is served. Leaders must be at the dinner table; they should attend or at least listen to how the dinner went. Many scientists avoid safety training because it is torture to attend, poorly planned and executed, and as devoid of nutrition as junk food. If leaders attended the safety training every year, just as laboratory staff must, what is being served at the table would become evident, and there would probably be changes in the overall program.

Leaders and safety officials must be on the same page even when they disagree. If leadership, safety, and science are sitting at the table, the leader must observe what is happening. Maybe the safety official decides to demand that all scientists wear six pairs of gloves. The laboratory staff could respond that there is no literature that backs this demand up and that doing this will place a huge burden on time and cost to science, which is simply unacceptable. At this point, the leader only nods in response. During a break, the leader approaches the safety official privately and requests a different procedure: staff will wear only two pairs of gloves. The leader presents this as his or her decision and requests that safety make it known moving forward that is what the organization is expecting. Disagreements between safety and leadership are to be handled privately, so that trust and respect are maintained among all three members of the family.

5. LEADERS ESTABLISH BALANCED REPORTING STRUCTURES

Leaders must serve as the balance between science and safety. One of the biggest challenges for a leader is establishing a reporting structure that controls for bias between science and safety. My consulting experiences have led me to conclude that one of the biggest barriers to a culture of safety is a poor reporting structure. How do leaders foster a sense of family, build trust, and empower those they are leading? It starts with a sound reporting structure with simple expectations.

There is a very popular leadership success story about the CEO at Alcoa. The highlights of the story are that Alcoa was not doing well. The trend with most organizations is that when leadership is not performing up to expectation, you replace them. That is what Alcoa did. As a publicly traded company, Alcoa introduced the new CEO, Paul O'Neill, who was asked to present to the shareholders. I believe the new CEO knew what he was about to do, but I don't think he could have imagined the multiple impacts that his first choice as CEO would have on the company.

O'Neill stated three expectations: (i) that all members of the organization, from leadership to the workforce, commit themselves to safety; (ii) that if an accident occurred, it would be reported to the CEO within 24 hours; and (iii) that when the accident was reported, a plan to ensure that it didn't occur again should accompany the accident report. Shareholders are used to hearing about ways to increase profits, not safety. By focusing on worker safety and expecting these three things, O'Neill was able to address the multiple challenges leaders face in establishing a balanced reporting structure. During his tenure, he also increased Alcoa's profitability.

The first challenge leaders run into with reporting structures is that they are too complex, with arrows going here, there, and everywhere. Like the example above, a reporting structure should be very simple. Safety is not only for those at the lower levels of the organization; instead, the highest level of leadership should be involved in the safety efforts of the organization. Therefore, the very first requirement is having the safety official report to the leader of the organization. Nothing should stand in the way of safety and leadership. A true safety culture could be established if all CEOs (private), presidents (academic), and directors (federal) were directly connected to the safety initiatives within their organizations.

This is where we run into the second challenge of reporting structures: the unconscious incompetence of believing that you, the leader, know the safety issues but not really knowing what you don't know. Many leaders will say they have a direct safety report. I ask them, "What are the safety issues within the laboratory? When was your last accident? What was done to ensure that it does not happen again?" Some argue they could not possibly afford to be involved at that detailed a level, but leaders who make the choice to know can foster a culture that goes well beyond safety, as Paul O'Neill did. When a leader engages directly in safety, a culture of safety is born. For parents, the safety of their home and family is of highest priority. If the leader of an organization can tell me where the priority of safety stands, I can, with a high level of reliability, predict the overall culture of safety within that organization.

A third challenge for reporting structures within organizations are federal regulations. These regulations call for organizational officials to become regulatory. This is equivalent to taking a stranger off the street, placing him or her in a family, and then asking asking him or her to be more loyal to the regulating body than to the family he or she belongs to and serves. How a leader controls for regulatory compliance and sustains a culture of safety is a challenge. Although leaders can foster an environment of trust and transparency, regulators are bound by law to report unexpected occurrences, which then lead to punitive actions that will certainly diminish a culture of safety within any organization.

Unfortunately, safety is diluted in many ways, leading to multiple barriers between leadership and the safety of the workforce. All accidents should be reported directly to the highest level of leadership. Paul O'Neill's three expectations made him a successful CEO, but it is rare to find organizations in which these expectations are highlighted.

6. LEADERS SEE FAILURE AS AN OPPORTUNITY TO IMPROVE

There are people who call themselves leaders but choose to blame everyone else when something goes wrong. Responsible leaders don't pass the blame for failure, they own it.

This book focuses on behavior and safety. So, "failure" in the context of this book is failure to behave in a desired way. Many rush to blame the individual who is not behaving correctly but fail to ensure that the individual was prepared to behave in the first place. For example, if the SOP is written in English and the individual does not read or is not fluent in English, how can that person be expected to follow the SOP? Leaders look at failure as an opportunity to learn and become better. But to learn from it, you must own it, and rather than pointing at someone else, leaders just point to something that can be improved and challenge themselves to improve.

In prescribing safe behaviors, we are asking people with different backgrounds, experiences, and education levels to behave in the same or similar fashions. Throw in the fact that we are humans and that the environment and those behaving around us directly impact our behavior, and you understand that leaders have much to do to ensure that behavior of all staff occurs in a proper fashion. When someone fails to behave as desired, leaders own it, so they can learn from it.

7. LEADERS SHARE SUCCESS WITH OTHERS

I have often wondered why, in life science publications, most scientists choose not to recognize the entire team for the success of the research project. Scientists certainly deserve credit for the research findings, but where is the recognition for the safety officer, animal care staff, engineering team, and whoever else participated in keeping what was needed successfully running? Leaders understand that success requires a team effort. When success is achieved, all those who have contributed should be recognized for their contributions.

8. LEADERS WELCOME DISAGREEMENT

There is no doubt that people like to be around those who agree with them. When we find someone who agrees with us, rarely challenges us, and supports us, we get things done and the experience is often pleasurable. However, true leaders understand that this scenario is nothing more than a bubble, and although much can get done, the lack of varying perspectives and opinions challenges the overall quality of the work.

Leaders should expect and seek challenging voices. Agreement is not a determinant of success for a leader; it's quite the contrary. When I witness leaders who are challenged by those they serve, it makes a profound statement about the leader and the organization. In the end, the leader will decide, but that is not what matters most here. What matters most for the leaders is that this practice encourages integrity and eliminates any inequity about the value of their thoughts versus those of the workforce.

9. LEADERS FOCUS ON WHAT WENT WRONG, NOT WHO

Leaders know that a feature of being human is imperfection. No matter how hard staff try, mistakes and accidents will happen. Success does not depend on staff skill alone. Laboratory staff are constantly interacting with environmental conditions that are not static and change every day, sometimes many times a day. Additionally, staff are in constant need of the resources required to behave in the desired fashion. When we put skills, resources, and environmental changes together, it is a recipe for an unexpected occurrence, i.e., an incident, accident, or near miss.

Unfortunately, many organizations have a culture of focusing on *who* went wrong rather than on *what* went wrong, playing the blame game. An incident investigation can be focused on finding who is at fault for something that may have been the result of many failures, not just one. It is a major mistake to assume that an incident occurred because of a single reason.

That approach is a very shallow and incomplete way of looking at unexpected occurrences.

If leaders focus on who went wrong, their efforts impact only one person directly, the one who is blamed for the unexpected occurrence. If leaders punish this person for the unexpected occurrence, this can have a profound impact on the level of transparency of future unexpected occurrences. Let's think about it from the perspective of the staff: I have a job because I must provide for myself and family. This job also provides a sense of self-value and belongingness for me. When an accident happens, you want me to report something that you are going to say is a mistake on my part? You want me to report a situation that could lead to professional and personal embarrassment, tarnish my reputation, or even cause me to lose my job?

You cannot expect laboratory staff to report unexpected occurrences if leadership is punitive. If a leader decides to focus on who went wrong, even in a manner that does not lead to professional scrutiny or loss of opportunity, the fact that you singled them out for an incident, accident, or near miss is punishment enough.

Focusing on what went wrong begins by investigating the sequence of events leading up to the unintended occurrence. Leaders recognize that for this one incident to occur, there had to have been multiple failures leading up to it. If leaders focus on what can be done to prevent future failures from occurring, they shift the blame away from who went wrong and address what went wrong. Fixing what went wrong fixes the safety issue for everyone.

For an effective safety culture to exist, leaders must foster transparency between themselves and the workforce. A shift away from the blame game is one of many steps needed to foster the trust between leadership and laboratory staff that must be present for a safety culture to exist.

10. LEADERS CARE

Just like a good coach or parent, leaders care about their teams and families. I created the acronym CARE because it encompasses the true essence of leadership. Leaders demonstrate CARE when they ensure compliance by applying accountability after providing resources and establishing expectations. The first two actions of CARE are establishing expectations and providing the resources needed to live up to those expectations. Leaders must start here when demonstrating CARE.

Establishing Expectations

Leaders must provide the workforce with expectations. Whether it is in the laboratory or the health care setting, expectations are typically provided within specific SOPs. As I've explained, SOPs are really standardized behaviors that leaders expect the workforce to follow so that a reproducible safety outcome occurs regardless of who is behaving. Keep in mind, people are not robots, and they have different levels of education, experience, and competence. Managing risks effectively means developing an evaluated and validated plan that adequately controls the general risk. This plan becomes an expectation, and once an expectation is established, leaders must ensure the expectation can be met.

Providing Resources

One of the reasons an expectation is not met is because those being asked to live up to the expectation don't have the resources to do so. In chapter 8, I discuss the items that are needed for sustained behavioral practices. Because expectations are behaviors, leaders must ensure that staff have what they need to behave as expected. Needs can be established when they understand the risk they face and the benefits they gain specific to the expectation. Once an explanation has been provided, staff must gain access to resources needed to live up to expectation. However, once a resource has been provided, it is an enormous risk to assume that staff have the skills needed to utilize the resource properly. Therefore, leaders must ensure skills training occurs and a sense of self-efficacy ("I can do it!") is established before asking staff to live up to any proposed expectation.

Ensuring Compliance

If an expectation is established, and nobody cares whether the expectation is being met, does the expectation truly exist? Ensuring compliance is an effort of observation. During observation, leaders can determine if someone is trying to meet, is unaware of, or is blatantly ignoring existing expectations. Measuring compliance determines one's effort in attempting to live up to expectations that leaders have implemented to keep the workforce safe. If leaders don't gauge the efforts their workforce is making to comply with expectations, it's as though those expectations don't exist. Because living up to expectations requires energy, attention, resources, and time, it's foolish to assume that humans are complying; you must observe them.

Applying Accountability

Accountability is the response leaders make to their observations regarding compliance. Unfortunately, accountability gets a bad name. There are two types of accountability. Leaders must exhibit both types equally, although with different responses.

Positive accountability is letting the workforce know that they are living up to or exceeding leadership expectations. Letting people know when they are doing something right not only solidifies the expected behavior but also fosters trust and sustains the relationships needed when something does go wrong in the future. Leaders recognize those who are living up to expectations.

Negative accountability is addressing those in the workforce who are failing to live up to expectation. There are varying degrees of negative accountability. For example, you would not treat someone who is trying to live up to expectation the same way as someone who is insubordinate and is choosing to fail to meet expectations.

When someone is trying and failing to meet expectations, leaders must once again ensure that this person has what is needed to sustain the expected behavioral practices. In reviewing the situation, leaders will typically find that something had not been provided and the individual has specific unmet needs that prevented expectations from being achieved. Perhaps he or she is unaware of the expectations or lacks the proper knowledge or equipment.

However, the biggest challenge is when someone chooses to fail to meet leadership expectations. This is like a member of a team telling a coach they won't practice or a child ignoring a parental request. It is how that coach or parent chooses to respond when this behavior occurs that not only sets the tone for the future of the player or child but also for the team and the family. If a leader fails to address intentional dismissal of leadership expectations, it gives that person permission to challenge any organizational expectation that exists. Furthermore, it enables those around the worker to begin modeling this mentality, which produces additional insubordination and a safety culture problem. Leaders must act to correct members of the workforce who fail to meet established expectations, and they should do so in a consistent, immediate, and clear fashion.

I have never encountered people who want to come to work and intentionally hurt themselves or others. Safety is an inherent part of science and health care. However, what people fail to understand is that they don't live in a bubble. Their individual behavior has an impact on the safety of others and themselves. Leaders remind them of this, while controlling for the human risk factors that naturally exist. In short, leaders CARE about those they serve.

BIOSAFETY *in the First Person*

The Road Less Traveled into Biosafety by Bob Ellis

Bob Ellis gets the last word in the book for many reasons. He first introduced me to "cowboy ethics," a term he coined and brought into the profession of biosafety. Bob doesn't just talk about these things; he rides for the brand, takes pride in his work, does what must be done, is tough but fair, and talks less while saying more. I will also be eternally grateful for Bob because of a phone call he made one day as I was preparing to serve the nurses and doctors who were treating the first two cases of Ebola in the United States. Bob's words were to the point. In short, he told me that he thought I was one of the most capable, the best, at what

I do, and that he knew I could do this job and do it well. There are few times when people say exactly the right thing at the right time. Bob did, and for that I remain grateful. His words gave me the assurance that I could serve, and serve I did. Thank you, Bob, for being a leader and an asset to the profession of biosafety.

Let's start with a question: How do most biosafety professionals get into the profession? In countless conversations with colleagues, I have heard that most did not start with the goal of being a biosafety professional. That has changed for the better in the past decade, as there are more opportunities for biosafety professional training, but in previous decades the training opportunities were very limited. Most people entered the profession by being appointed to the biosafety officer position as a responsibility added on to their other faculty or research leader positions. This was the path of my initial foray into the biosafety profession. I was appointed to the University Biohazard Committee in August 1978, my first year at Colorado State University (CSU). I enjoyed the opportunity to review the rDNA and infectious disease research at our university. We wrote our first university biohazard handbook in the early 1980s. It was in 1985, when I was chair of the Biosafety Committee (whose name had been changed from Biohazard Committee because "safety" sounded better than "hazard"), that we determined that we needed a university biosafety officer. Two of us took the request for funding to our vice president for research. He listened carefully to our request for funding three months per year for the position, to be filled internally with the successful CSU applicant, and asked: "Why do we need a biosafety officer? No one has died, have they?" Our answer was a quick, "No one has died, and we want to keep it that way." We were granted the funding. I applied for the position and became CSU's first biosafety officer in January 1986. I enjoyed the position and found that our university was amenable to working with me to ensure that our research was conducted safely.

During the summer of 1989, I was approached by the director of our Veterinary Diagnostic Laboratory asking if I was interested in leading the bacteriology section of the lab. I thought about the opportunity, the responsibility, and my previous experiences in a veterinary diagnostic lab prior to CSU. I accepted the challenge and resigned as the university's biosafety officer in October 1989. I enjoyed the diagnostic lab responsibilities as well as my continued teaching and research. In June of 1992, in a visit with our dean of the College of Veterinary Medicine and Biomedical Sciences, I was told that my leadership as the bacteriology section head in the diagnostic lab was no longer needed. Wow: for the first, and only, time in my life in any position, I was fired!

Then a series of events preceded my reentry into biosafety. This is where the "road less traveled" back to biosafety began to stretch ahead of me. It took five years, during which I was very busy with my teaching and research, before the biosafety officer position at CSU was open again. A national search was announced, and after much thought and contemplation, I applied. I was again the successful candidate, with one large caveat. The position was advertised as a full-time position, and I wanted it as only a half-time position, because I still had my teaching, research, graduate students, etc. Even though I was the top candidate, I had to wait for two other candidates to turn down the offer before it was offered to me, and I immediately accepted. In short, I had gone from "You are fired" to stepping deeply into a lifelong and rewarding career in biosafety. The last 20+ years have been very rewarding from the standpoint that we have developed a nationally recognized biosafety program at CSU. I have not done this alone, by any stretch of the imagination; there have been countless colleagues who have given honest and frank advice through the years. These colleagues have been inspirations to keep building, keep improving, and keep on an upward path so that our biosafety program will always be one that garners, first of all, local respect and buy-in, and then respect from colleagues outside our university.

The colleagues who have helped shape our biosafety program and me as a trusted biosafety professional have been in many walks of life, not exclusively in biosafety. I try to maintain a constant vigil over those around me, no matter their position in relation to biosafety. One of the most important lessons I learned was from a colleague in Lima, Peru. I had been in Peru many times as part of a USAID research project, working with small-ruminant diseases in the sheep and alpaca populations in the very high mountains of Peru. When in Lima, I had noticed an elderly couple at a street corner. The man would ask for money when the traffic was stopped. My friend and I were driving to a meeting, and he asked me, when we were at that corner, if I had noticed this couple. I said yes, I had, every time I was in Lima. My friend said that a few days ago, he had stopped for the light and was approached by the man for a handout. He told him he could not take time to give him anything, he was in a hurry, and to not bother him. The man calmly told my friend, "I have asked you politely. You can answer 'no' politely. I am very poor, but I have my dignity. Please do not try to take my dignity from me." I have never forgotten this encounter, and in my dealings with anyone—here at the university, in daily life, whoever the person is—I try to remember that we all have our dignity and to deal with others so that they know we are not attempting to take their dignity from them. I believe that this has assisted me in dealing with others and in establishing communication where we can all learn to listen, discuss, and then act in a manner that is best for us all.

I am very thankful that I was granted another opportunity to be in the biosafety profession, and I am thankful for the wealth of opportunities that have been available to explore the world of biosafety and biosecurity over the years since 1997, and even a few years before that. In conclusion, there is no way that we accomplish our goals alone. Many, many people whom I consider dear friends and colleagues, many others who are acquaintances, and still others whose writings I read or whose teaching I watch have contributed to my continued growth in the biosafety profession. I thank all these people, and most especially my caring and supporting partner and wife, for all they have done for me in the journey along the road less traveled. There are still many miles and years to go! And please keep being the "wind beneath my wings," as I will keep trying to be the wind beneath your wings.

Image Credits

Page 2, top: Demonstration of Ebola training in field, cleaning up body fluids. Source: Dr. Ellen Dotson, CDC18702; bottom: Ebola testing in the field. Source: John Saindon, CDC22515

Page 8, top: Men in PPE in ambulance. Source: Chaikom©shutterstock.com, 228517582; **bottom:** Woman working in lab. Source: MihasLi©shutterstock.com, 391395604

Page 16, top: Taking blood in Africa. Source: Leonie Broekstra ©shutterstock .com, 251585581; **bottom:** Washing a petri dish. Source: Rattiya Thongdumhyu©shutterstock.com, 609512942

Page 26, top: Woman in mask. Source: Gorodenkoff©shutterstock.com, 691541104; **bottom:** Woman pipetting in biosafety cabinet. Source: KYTan©shutterstock.com, 391395604

Page 40, top: Petri dish with organisms. Source: KuLouKu©shutterstock.com, 663446374; **bottom:** Biohazard bag. Source: Sherry Yates Young©shutterstock .com, 278509241

Page 64, top: Man working in lab. Source: lightpoet©shutterstock.com, 139507928; **bottom:** Woman reading in lab. Source: anyaivanova©shutterstock.com, 537098116

Page 76, top: Donning PPE in the field. Source: Athalia Christie©shutterstock.com, CDC17840; **bottom:** Working in safety cabinet. Source: CDC, CDC19962

Page 90, top: Lab technicians and computer. Source: Leonardo da©shutterstock .com, 445348096; **bottom:** Lab technicians with microscope. Source: ASDF_ MEDIA©shutterstock.com, 1146229706

Page 102, top: Teacher and student viewing slide. Source: Dmytro Zinkevych©shutterstock.com, 763892554; **bottom:** Lab technician in biosafety laboratory. Source: Pressmaster©shutterstock.com, 497486887

Page 117, Figure 9.1. Source: Sean G. Kaufman

Page 122, top: Discussion in laboratory. Source: Pressmaster©shutterstock.com, 475109248; **bottom:** Placing something in safety cabinet. Source: Tonhom1009©shutterstock.com, 403782517

Page 142, top: Donning PPE in the field. Source: U.S. DoD, Army Sgt. 1st Class Tyrone C. Marshall, Jr., CDC18352; **bottom:** Emergency team exercise at airport. Source: Onalee Grady-Erickson, CDC23095

Page 158, Figure 11.1. Source: Sean G. Kaufman

Page 164, top: Working in safety cabinet. Source: Tonhom1009©shutterstock.com, 408838903; **bottom:** Working in PPE in the field. Source: Tara Sealy, CDC20974

Page 167, Figure 12.1. Source: CDC

Page 180, top: Woman being instructed on Ebola testing. Source: CDC/ Alaine Knipes, CDC22504; **bottom:** Training in Haiti using video. Source: Anita D. Sircar, CDC22784

Page 182, Figure 13.1. Source: Elizabeth R. Griffin Research Foundation

Page 192, top and bottom. Source: Sean G. Kaufman

Page 204, top: Filling out safety form. Source: Micolas©shutterstock.com, 250910614; **bottom:** Surveillance camera with smartphone. Source: poylock19©shutterstock.com, 704092561

Page 220, top: Teacher and students in lab. Source: Zhu difeng©shutterstock.com, 384646627; **bottom:** Preparation for airport screening for Ebola. Source: CDC/ Sally Ezra, CDC17902

Biohazard symbol, source: nexusby©shutterstock.com

Index

A

Accountability, 47–48
  leaders applying, 231–232
  programs, 80
Acyclovir, 6
Administrative controls
  accountability, 47–48
  compliance, 47
  fit-for-duty screenings, 45–46
  incident surveillance, 46
  medical surveillance, 46
  training, 48–49
  vaccination, 45
Agnes Scott College, 61
Agricultural Research Service (ARS), USDA, 202
Alderman, Lee, 87–89, 147, 193
American Association for Laboratory Animal Science, 202
American Biological Safety Association (ABSA), 36, 62, 137, 197–203
American Biological Safety Association International, 179, 197, 203
American Media, Inc. (AMI), 137, 138
American Society for Microbiology, 71, 198, 213
American Society of Association Executives (ASAE), 202–203
Anglewicz, Carrie, 22–25
Animal biosafety levels (ABSL), 11
Applied Biosafety (journal), 198
Applied Laboratory Emergency Response Training (ALERT), 146–150
  assessment, 148, 149
  risk perception evaluation, 148, 149
Ascending disseminated myelitis, 6
Association of Primate Veterinarians, 61–62
Association of Public Health Laboratories, 202, 218–219

Attitude, behavior, 67
Augustine, James, 147
Authority, 28
Automated external defibrillator (AED), 157

B

Bacillus anthracis, 99
Bacillus subtilis, 99
  biosafety level 1 (BSL1), 10
  clinical containment level 1 (CCL1), 13
Bade, John, 35
Barbeito, Manny, 36, 199
Barkley, Emmett, 36
Beaking method
  glove removal, 193, 197
  step-by-step demonstration, 194–196
Behavior
  abilities, 95
  access to resources, 94–95
  complacency, 31
  learning a new, 30
  perceived mastery of, 30–31
  self-efficacy, 95–96
  sustained, 93–96
  understanding benefit, 94
  understanding risk, 94
Behavioral expectation contract, 116, 117
Berkelman, Ruth, 103
Biological agent-human interface, risk factor, 31–32
Biological laboratory containment, 9–11
Biological risk mitigation
  evolution of, 18–22
  hazard identification era, 18–19
  phases of, 17–18
  risk assessment era, 19–20
  risk communication era, 21–22
  risk management era, 20–21

Biological risks, emergency preparedness and
    response to, 150–159
Biological Safety, 5th edition (ASM Press), 216
Biosafety and Biosecurity Committee, APHL, 219
Biosafety cabinets, 215–216
*Biosafety in Microbiological and Biomedical
    Research Laboratories* (CDC/NIH)
    (BMBL), 36, 41–44, 54, 73, 121, 127
Biosafety in the first person
    Alderman, Lee, 87–89
    Anglewicz, Carrie, 22–25
    Byers, Karen, 190–191
    Cappola-Vojak, Dottie, 13–15
    Ellis, Bob, 232–235
    Griffin, Caryl, 5–7
    Hawley, Robert, 137–140
    Kanabrocki, Joseph, 34–39
    Kozlovac, Joe, 174–179
    Mathews, Henry, 72–75
    Pentella, Mike, 212–219
    Stygar, Ed, 197–203
    Trevan, Tim, 159–163
    Troiano, Anthony (AJ), 97–100
    Welch, Jim, 60–62
    Ziegler, Sarah, 118–121
Biosafety level 1 (BSL1), 9, 10, 33, 43
Biosafety level 2 (BSL2), 9, 10, 33, 44, 134, 216
Biosafety level 3 (BSL3), 9, 10, 33, 42, 103,
    150, 175
Biosafety level 4 (BSL4), 9–10, 33, 103, 150,
    151, 156
Biosafety officer (BSO), 35–36
Biosafety training, 65
Bird flu, 202
Blame game, incidents, 209–210
Blink (Gladwell), 17
Bloodborne Pathogen Standard, 73, 74
BMBL. *See Biosafety in Microbiological and
    Biomedical Research Laboratories*
    (CDC/NIH)
Booties, 53
Boston University, 147
Brantly, Kent, 97, 154
Brucella, 190, 215, 217–218
Buckley, Mary, 198, 200
Bureau of Laboratories, Michigan, 22
Burke, Paul, 150–151
Burnett, LouAnn, 199, 203
Byers, Karen, 190–191, 198, 201, 203

C

Cadler School of Theology, 61
Called for Life (Brantly), 154
Campus Safety, Health, and Environmental
    Management Association, 202
Cappola-Vojak, Dottie, 13–15
Cardiopulmonary resuscitation (CPR), 156–159

CARE acronym
    accountability, 231–232
    compliance, 231
    expectations, 230–231
    resources, 231
Casadaban, Malcolm, 34, 36–39, 45, 133, 207
Cat scratch fever, 6
Center for Global Health Science and Security,
    Georgetown University, 7
Centers for Disease Control and Prevention
    (CDC), 3, 5, 24, 38, 41, 73, 202, 217
    behaviors for infectious disease outbreak, 145
    Epidemiology and Laboratory Capacity
        grant, 219
    protocol for removing PPE, 166, 167, 169
    Technical Development Laboratory
        (CDC/TDL), 87–88
Certified Biological Safety Professional, 198
Chicago Department of Health, 38
Chicago Department of Public Health
    (CDPH), 38
Chicago Police Department, 38
Chlamydia trachomatis, 213
Cipriano, Mary Ann, 199
Climate, 123
Clinical containment, 11–13
Clinical containment level 1 (CCL1), 12, 13
Clinical containment level 2 (CCL2), 12, 13
Clinical containment level 3 (CCL3), 12, 13
Clinical containment level 4 (CCL4), 12, 13
Clinical lab, 213–216
    biosafety cabinets, 215–216
    lab-acquired infection, 214–215
    staff training in biosafety, 216
    tuberculosis exposure, 215
    vaccines for staff, 214
    See also Laboratory experience
Clinton Administration, 177
Clostridioides difficile, 13, 99
Coats
    laboratory and health care provider, 53–54
    sequence for removing, 167
Colbert, Stephen, 160
College of American Pathologists, 216
College of Veterinary Medicine and Biomedical
    Sciences, 233
Colorado State University (CSU), 233–234
Complacency, 84–85
Compliance, 47, 231
Condreay, Patrick, 203
Congo-Crimean hemorrhagic fever virus
    biosafety level 4, 10
    clinical containment level 4 (CCL4), 13
Conjunctivitis (pink eye), 6
Containment
    biological laboratory, 9–11
    clinical, 11–13
    process of, 77

Containment philosophy, 77
  agents, 78–83
  engineering controls, 83
  leadership controls, 78–80
  organization, 85–87
  people, 83–85
  personal protective equipment (PPE), 82–83
  standard operating procedures (SOPs), 80–82
Continuous improvement, ten commandments
    of, 162–163
Coolidge, Calvin, 176
Coxiella burnetii, biosafety level 3, 10
Culture, 123
  One Health, 123
  One-Safe, 123–124
  See also One-Safe culture

D

David J. Sencer Museum, 50
DECON (decontamination) procedure, 137–139
Department of Defense (DOD), 36
DeServi, KariAnn, 201, 203
Dierickx, Ingemar, 160
Downing, Marian, 203
Doxycycline, 6
Duane, Elizabeth Gilman, 200, 201

E

Eagleson Institute, 202
Ebola virus, 119
  agent-human interface, 31
  biosafety level, 10
  cases in United States, 96–97, 144, 218–219,
    232–233
  clinical containment level 4 (CCL4), 13
  Emory Hospital treating patients, 104
  first person in United States with, 97
  medical surveillance programs, 208
  outbreak (2014), 3, 50, 121, 133, 166, 193, 205
Elizabeth R. Griffin Foundation, 7, 181, 182,
    201–202
Ellis, Maureen, 201
Ellis, Robert, 201, 232–235
Emergency
  definition, 144
  mitigation phase, 146
  preparedness phase, 144
  recovery phase, 145–146
  response phase, 144–145
  unconscious individuals, 156–159
  See also Applied Laboratory Emergency
    Response Training (ALERT)
Emergency preparedness and response to
    biological risks, 150–159
  evacuations, 150–152

  gross contamination, 155–156, 157
  needlesticks and eye splashes, 152–154
  spills, 154–155
  unconscious individuals, 156–159
Emory Healthcare Isolation Unit, 131, 209
Emory University, 61, 89, 103, 144, 172,
    193, 201
Emory University Healthcare, 4, 50, 96
Emory University Hospital, 104, 154
Emory University Science and Safety Training
    Program, 52
Emotional state, risk factor, 33–34
Engineering controls
  building, 49
  devices, 50
  rating program, 57, 58
  See also Primary controls of safety
Environment, risk and standard operating
    procedures, 28–29
Environmental Health and Safety Office, 89
Escherichia coli O157, 190
Escherichia coli O157:H7, 3
European Biological Safety Association, 202
Evacuations
  emergency response, 150–152
  green, 151
  red, 152
  yellow, 151
Experts, training, 183–184
Eye protection, 54
Eye splashes, 152–154

F

Face shield, sequence for removing, 167
Federal Select Agent Program, 115
52 Weeks of Biosafety (program), 71
Fink, Richie, 199
First, Melvin, 198, 199
First Biosafety and Biosecurity International
    Conference, 161
First person accounts. *See* Biosafety in the
    first person
Fit-for-duty screenings, 45–46
Fleming, Diane, 199
Fontes, Ben, 201
Francisella tularensis, 131
Funk, Glenn, 201

G

Ganciclovir, 6
Gandhi, 113, 185
Gehrke, Michelle, 203
George Mason University, 150
Georgetown University, 7
Gervais, Ricky, 160
Gilchrist, Mary, 216

Gillum, David, 198
Gilpin, Richard, 175
Global Health Security Agenda Consortium, 202
GloGerm, 92, 156, 166, 169, 172–173, 193
Gloves, 51
  beaking method for removal, 193–197
  sequence for removing, 167
  step-by-step removal demonstration, 194–196
Goggles, sequence for removing, 167
Great Lakes Center of Excellence for Biodefense and Emerging Infectious Diseases Research, 35–36
Green evacuation, 151
Griffin, Beth, 3, 5–7, 60–62, 89, 130, 201, 207
Griffin, Bill, 201
Griffin, Caryl, 5–7, 201
Gupta, Sanjay, 169
Gurtler, Joshua, 3

H
Harding, Lynn, 199
Harran, Patrick, 131
Hawley, Robert, 36, 137–139, 178, 201
Hazard identification, 18–19
Health surveillance programs (HSPs), 206–208
  continuous HSPs, 207–208
  initial HSPs, 206–207
  reactive HSPs, 207
Hemochromatosis, 38
HEPA (high-efficiency particulate air) filters, 42, 49, 83, 88, 139
Hepatitis A virus (HAV), 213
Hepatitis B virus (HBV), 213, 214
  biosafety level 2, 10
  clinical containment level 2 (CCL2), 13
Heroism, 4
Herpes B virus, 6, 88, 201
High-Containment Laboratory Accreditation program, 201
HIV (human immunodeficiency virus), 75, 133, 175
  biosafety level 2, 10
  clinical containment level 2 (CCL2), 13
  clinical lab, 213, 214
  patient with end-stage HIV disease, 14–15
Homovec, Bill, 201
Human behavior, 65–66
  person or environment, 67–68
  punishment versus reinforcement, 70–72
  safety culture, 67
  understanding, 66–67, 68–70
Human risk factors
  appealing to authority, 28
  biological agent-human interface, 31–32
    emotional state, 33–34
    environmental conditions, 28–29
    individual capability, 33
    mental state, 30–31
    physical state, 32
    Swiss cheese effect, 27–28
Hunt, Debra, 199

I
Illinois Department of Health, 38
Incident surveillance programs, 46, 209–210
  blame game, 209–210
  punishing reporters, 210
  reporting process, 209
Individual capability, risk factor, 33
Infection control, clinical containment, 11–13
Infectious disease pioneers, 3–5
Influenza virus, clinical containment level 2 (CCL2), 13
Intrinsic motivation
  lots of "I don't know"s, 112–113
  lots of silence, 111–112
  lots of urges to jump in, 113
Intrinsic safety, 103–104
  behavioral expectation contract, 116, 117
  extrinsic motivation, 104
  intrinsic motivation, 105–117
  program examples, 113–117
  reviewing standard operating procedures, 114
  safety audit facilitation by workforce, 115
  serving rather than policing, 114–115
  systemic motivation, 104–105
  workforce presenting to safety committee, 116
  See also Motivation
Inventory surveillance programs, 211–212
Iowa State Hygienic Laboratory, 216
Irwin, Steve, 20

J
Jenkins, Brenda, 88
Johns Hopkins Institute, 177
Johns Hopkins University, 175
Johnson, Barbara, 198, 199
Johnson, Diane, 203

K
Kanabrocki, Joseph, 34–39
Kaufman, Sean, 89, 100, 201
Keene, Jack, 199
Knudsen, Richard, 198
Kornberg, Arthur, 99
Kozlovac, Joseph, 174–179, 201, 202

L

Laboratory experience
  biosafety cabinets, 215–216
  clinical lab, 213–216
  competencies and capabilities, 218
  early years, 212–213
  Ebola in the United States, 218–219
  establishing biosafety officer position, 219
  lab-acquired infection, 214–215
  public health laboratory, 216–218
  staff training in biosafety, 216
  tuberculosis exposures, 215
  vaccines for staff, 214
Lansing State Journal (newspaper), 24
Leadership. *See* Responsible leadership
Leadership controls, 43, 45–49
  accountability, 47–48
  accountability programs, 80
  compliance, 47
  containment philosophy, 78–80
  fit-for-duty screenings, 45–46
  incident surveillance, 46
  medical surveillance, 46
  rating program, 58, 59
  surveillance programs, 78–79
  training, 48–49
  training programs, 79–80
  vaccination, 45
  See also Administrative controls; Primary
      controls of safety
Leadership expectations
  leaders caring, 129–130
  leaders preparing workforce, 127–128
  leaders promoting safety, 129
  leaders protecting the workforce,
      128–129
  One-Safe culture, 127–130
Leadership Institute for Biosafety Professionals,
    21, 201

M

McClure, Harold, 89
Marburg virus, biosafety level 4, 10
Masks, surgical, 52–53, 167
Maslow, Abraham, 29, 66
Mathews, Henry, 3, 50, 72–75, 147, 166,
    193
Maybelline (delivery nurse), 3, 205
Medical surveillance, 46, 79
Medical surveillance programs, 208–209
Meechan, Paul, 203
Mental state, risk factor, 30–31
Methicillin-resistant *Staphylococcus aureus*
    (MRSA), clinical containment level 2
    (CCL2), 13
Meyer, Paul, 203
Michigan Department of Health Laboratory, 3

Middle East respiratory syndrome virus,
    clinical containment level 3
    (CCL3), 13
Midwest Regional Center of Excellence for
    Biodefense and Emerging Infectious
    Disease Research, 35
Milgram, Stanley, 28
Minarik, Barb, 200
Mitigation phase, emergency, 146
MMWR (journal), 25, 191, 216, 218
Morbidity and Mortality Weekly Report (CDC).
    See MMWR (journal)
Morland, Melissa, 203
Motivation
  continuing to ask question, 110–111
  extrinsic, 104
  intrinsic, 105–117
  intrinsically motivating others, 106–111
  learners, 184
  serving others, 108–109
  signs of practicing, 111–113
  systemic, 104–105
  treating people as possessing solution,
      109–110
  treating people as sensible, 110
  treating people as skilled, 109
  See also Intrinsic safety
Mycobacterium tuberculosis
  biosafety level 3, 10
  clinical containment level 3 (CCL3), 13

N

National Biocontainment Training Program
    (NBBTP), 119
National Biosafety and Biocontainment
    Training Program, 178
National Cancer Institute, 175
National Emerging Infectious Diseases
    Laboratory (NEIDL), 150
National Institutes of Allergy and Infectious
    Diseases (NIAID), 35–36
National Institutes of Health (NIH), 35, 41, 73,
    103, 147, 202
National Registry of Certified Microbiologists,
    198
Needlesticks, 152–154
Negative-pressure respirators, 51–52
Neisseria, 22, 23
Neisseria meningitidis, 22, 24, 191, 215
Nellis, Barbara Fox, 203
Nigeria gruberi, biosafety level 1, 10
*NIH Guidelines for Research Involving
    Recombinant or Synthetic Nucleic
    Acids*, 36
Nonpathogenic *Escherichia coli*, clinical contain-
    ment level 1 (CCL1), 13
Novices, training, 183

O

Occupational Safety and Health Administration (OSHA), 24, 38, 115, 137, 177, 202, 214, 217
Ohio State University, 212
O'Leary, Maureen, 203
One Health, concept of, 123
O'Neill, Paul, 227–228
One-Safe culture, 123–124, 136
  common risks, 125
  common rituals, 126
  common rules, 125–126
  leadership expectations, 127–130
  safety official expectations, 134–136
  workforce expectations, 130–134
Organization, containment philosophy, 85–87
Outbreak (movie), 41, 118

P

Paramecium, 35
Paramecium tetraurelia, 35
Pentella, Mike, 212–219
People
  basic needs of, 83–84
  complacency of, 84–85
  humanity of, 85
Personal protective equipment (PPE), 11, 33, 42, 43, 50–54
  Bloodborne Pathogens Standard, 74–75
  comfort of, 82
  containment philosophy, 82–83
  eye protection, 54
  glove removal by beaking method, 193–197
  gloves, 51
  laboratory and health care provider coats, 53–54
  negative-pressure respirators, 51–52
  positive-pressure respirators, 52
  practicality of, 83
  rating program, 57–59
  selection of, 82
  sequence for removing, 166, 167
  shoe covers and booties, 53
  surgical masks, 52–53
  See also Primary controls of safety
Pham, Nina, 3, 100, 156, 166, 208
Physical state, risk factor, 32
Pioneer, 3
Plans
  applicable, 93
  learnable, 93
  necessary, 93
  practical, 92–93
  specific, 93
Plans plus behaviors equal outcomes (P + B = O), 91–92, 97
  desired outcomes, 96–97

  effective plans, 92–93
  sustained behaviors, 93–96
Plasmapheresis, 7
Positive-pressure respirators, 52
Practitioners, training, 183
Preparedness phase, emergency, 144
Primary controls of safety, 42, 43
  administrative, 43, 45–49
  engineering, 43, 49–50
  leadership, 43, 45–49
  levels, 42–45
  personal protective equipment (PPE), 43, 50–54
  rating your safety program, 57–59
  standard operating procedures (SOPs), 43, 54–57
Principles and Practices of Biosafety (course), 199
Prochaska stages-of-change model, 30
Pseudomonas taeniospiralis, 35
Public health laboratory staff, 4
Punishment, outcomes of, 29

R

Razin, Lena, 203
Rebar, Richard, 201
Recovery phase, emergency, 145–146
Red evacuation, 152
Reese, Linda, 3, 22–25, 45, 133, 207
Registered Biosafety Professional certification, 121, 198
Respirators
  MaxAir, 83
  negative-pressure, 51–52
  positive-pressure, 52
  powered air-purifying (PAPRs), 82, 83
Response phase, emergency, 144–145
Responsible leadership, 221–232
  applying accountability, 231–232
  caring, 230–232
  ensuring compliance, 231
  establishing balanced reporting, 227–228
  establishing expectations, 230–231
  fairness in, 223–224
  focusing on what went wrong, not who, 229–230
  preparing staff, 222–223
  promoting science and safety, 226–227
  protecting staff, 224–226
  providing resources, 231
  seeing failure as improvement opportunity, 228–229
  sharing success, 229
  welcoming disagreement, 229
Rhode Island Hospital, 98
Rhode Island State Health Laboratories, 100
Richmond, Jonathan, 199
Risk, understanding of, 94
Risk assessment, 19–20

Risk communication, 21–22
Risk factors. *See* Human risk factors
Risk management, 20–21
Risk mitigation. *See* Biological risk mitigation

S

Saccharomyces, 35
Safety
   definition, 41
   guidelines, 41–42
   levels of, 42–45
   practicing intrinsic motivation, 111–113
   rating your program, 44, 57–59
   serving others, 108–109
   See also Intrinsic safety; Primary controls
      of safety
Safety audits, workforce facilitating, 115
Safety committee, workforce presenting to, 116
Safety culture, 67, 77, 86–87, 104–105
Safety official expectations
   advocating for all, 136
   being seen, 135–136
   increasing capability, 135
   knowing standards of profession, 134–135
   serving rather than policing, 135
Safety surveillance programs, 206, 212
   health, 206–208
   incident, 209–210
   inventory, 211–212
   medical, 208–209
   video, 210–211
St. Louis encephalitis virus, biosafety level 3, 10
Salerno, Ren, 19
Salkin, Ira, 198
Salmonella
   biosafety level 2, 10
   clinical containment level 2 (CCL2), 13
   laboratory testing, 22, 23, 25
Sanchez, Anna, 55
Sandia National Laboratories, 202, 217
Sangji, Sheri, 131
Savage, Julie, 200, 203
Savage, Karen, 198, 200, 203
Schmidt, Jerome, 199
Science and Safety Training Program, Emory
      University, 172
Self-efficacy, learning behavior, 95–96
Setlow, Peter, 99
Severe acute respiratory syndrome, 13, 133, 202
Shigella, 190, 214
Shoe covers, 53
Spahn, Gerald, 199
Spills, 154–155
Standard operating behavior (SOB), 55, 165
Standard operating procedures (SOPs), 11, 20,
      21, 27, 43, 54–57
   biological agent-human interface, 31–32

containment philosophy, 80–82
   definition, 68, 165
   ensuring others follow, 131
   environment and, 28–29
   evaluation, 169–170, 172–173
   external validity, 171–172
   flexibility, 81–82
   following, 130–131
   gross contamination, 155–156, 157
   internal validity, 171
   purpose of, 168
   quality, 81
   quantity, 81
   rating program, 58, 59
   risk reduction in, 169
   safety official reviewing, 114
   spills, 154–155
   thoughts on, 173–174
   training, 184–188
   understanding, 169–170
   validation, 170–173
   verification, 173
   writing, 120, 168–169
   See also Primary controls of safety
Stygar, Ed, 197–203
Stygar, Edward J., Jr., 197
Stygar Associates, 197
Sullivan, Betty, 35
Surgical masks, 52–53
Surveillance programs
   leadership, 78–79
   See also Safety surveillance programs
Swiss cheese effect, risk factor, 27–28, 31, 32, 34

T

Tecker International, 203
Ten Commandments of Continuous
      Improvement, 162–163
Tepper, Byron, 36, 175, 199
Tetrahymena, 35
Thomas, John, 212
Thompson, Chris, 201
Toxoplasma, biosafety level 2, 10
Training, 48–49
   awareness session, 183
   education session, 183
   effective strategies, 181–190
   learners, 183–184, 189
   motivating learners, 184, 189
   Principles and Practices of Biosafety course, 199
   round 1, 181–183
   round 2, 184–186
   round 3, 186
   round 4, 186–187
   round 5, 187–188
   skills and abilities, 189, 190
Trevan, Tim, 159–163

Troiano, Anthony (AJ), 97–100
Troiano, Anthony J., Sr., 98
Tuberculosis, 213, 215
Tulis, Jerry, 199

U
Unconscious individuals, 156–159
Universidad Nacional Autónoma de
    Honduras, 55
University of California, Los Angeles, 131, 152
University of Chicago, 35–38
University of Connecticut, Health Center, 99
University of New Hampshire, 98, 99
University of Texas Medical Branch (UTMB),
    21, 118
University of Wisconsin at Madison
    (UW-Madison), 35
U.S. Army Medical Research Institute of
    Infectious Diseases, 36, 137, 139, 175
U.S. Biodefense Programs, Fort Detrick, 36
U.S. Department of Agriculture (USDA), 3,
    174, 202
U.S. Environmental Protection Agency
    (EPA), 137–139

V
Vaccination, 24, 45
Van Houten, Joe, 199
Veterinary Diagnostic Laboratory, 233
Video surveillance programs, 210–211
Vinson, Amber, 156, 166, 208

W
Wagener, Stefan, 200, 201
Washington University in St. Louis
    (WUSTL), 35
Welch, Jim, 60–62, 178, 201
West Nile virus, 216
White, Claude Wisdom, Sr., 177
Wilson, Debbie, 119
Workforce expectations
    following SOPs, 130–131
    One-Safe culture, 130–134
    reporting incidents, accidents, and
        near misses, 131–132
    reporting medical conditions,
        133–134
    reporting symptoms matching agent
        presentation, 132–133
World Health Organization, 41–42

Y
Yellow evacuation, 151
Yerkes National Primate Research Center, 3,
    5–6, 61, 89
Yersinia pestis CO92, 37–38
Yersinia pestis Kim D27, 37–38
Yersinia pseudotuberculosis, 37

Z
Ziegler, Sarah, 118–121
Zimmerman, Dee, 21